Communication and Bioethics
at the End of Life

Lori A. Roscoe · David P. Schenck

Communication
and Bioethics
at the End of Life

Real Cases, Real Dilemmas

 Springer

Lori A. Roscoe
Department of Communication
University of South Florida
Tampa, FL
USA

David P. Schenck
Morsani College of Medicine
University of South Florida
Tampa, FL
USA

ISBN 978-3-319-70919-2   ISBN 978-3-319-70920-8   (eBook)
https://doi.org/10.1007/978-3-319-70920-8

Library of Congress Control Number: 2017958596

Printed on acid-free paper

This Springer imprint is published by Springer Nature
The registered company is Springer International Publishing AG
The registered company address is: Gewerbestrasse 11, 6330 Cham, Switzerland

This book is dedicated to the people whose stories we share here (whose identities have been changed to protect their privacy). These stories belong not just to the patients, but also to the family members who faced unbearable situations where difficult decisions had to be made, and to the health care professionals, hospital administrators, and ethics committee members who agonized about how to do the best thing possible for each patient. We share these stories with the hope that we can all learn better ways to die and better ways to care for dying patients and their families.

Preface

This book is a collection of real-life cases exploring the complex range of issues inherent in contemporary end-of-life medical care. It is intended for physicians, medical students, residents, ethics committee members, social workers, chaplains, nurses, bioethicists, researchers, and scholars who confront ethical issues with patients and families at the end of life, and who are committed to an understanding of the ways in which things can go wrong in efforts to improve our American way of dying. Most Americans die in institutional settings, primarily hospitals, which involve a challenging set of circumstances to be considered in helping patients die well. Media saturation, concerns for privacy, institutional norms, cultural diversity, politics, technology, and advances in medical care all complicate the decision-making, communication, and ethical analysis that are part of the dying process. More individual concerns, including family dynamics, patient preferences, spirituality, and insufficient advance care planning, also confound solutions that satisfy all stakeholders.

End-of-life care is a controversial matter. The classic cases of In re Quinlan (1976) and Cruzan v. Director (1990) established the right of patients or their surrogates to refuse life-sustaining treatment, but the more recent case of Terri Schiavo (Caplan et al. 2006) demonstrated how difficult it can be to exercise this right in the face of family conflict, media coverage, and political chicanery. The same year that Schiavo lapsed into a persistent vegetative state, retired pathologist Jack Kevorkian euthanized his first patient in Oakland County Michigan (Roscoe et al. 2000). Kevorkian went on to assist the deaths of over 100 people illegally between 1990 and 1997 until his arrest, conviction, prison sentence, and eventual death. Today, 20% of Americans live in a state where physician-assisted suicide (which is a highly regulated version of what Kevorkian practiced) is legal; California is the most populous U.S. state to legalize this practice, and one of the most recent. Since Oregon passed its *Death with Dignity* statute in 1997, 1,545 terminally ill people have had lethal prescriptions written for them, and 991 have died from ingesting these medications (Oregon Public Health Division 2015). That same year, the U.S. Supreme Court heard two cases—Vacco v. Quill (1997) and Washington v. Glucksberg (1997)—and found no constitutionally protected right to die; the ruling was predicated in part by the hope expressed by Justice Sandra Day O'Connor that all Americans would be able to access high-quality end-of-life care.

The growth in the number of hospice programs in the U.S. increases the chances of receiving quality end-of-life care. This number has grown from fewer than 3000 in 1998 to over 5,800 currently, with the number of patients served increasing from 540,000 to 1,542,000 (NHPCO 2015). The median length of stay, however, has decreased slightly from 22 days in 1987, to just over 17 days in 2014 (Gage et al. 2000). Hospice care also remains an option overwhelmingly chosen by White patients; over 80% of hospice enrollees identify their race as White (NHPCO 2015). In 2006, the American Board of Medical Specialties (ABMS) officially recognized hospice and palliative medicine as a formal subspecialty of medicine in the United States focusing on symptom management, pain relief, and end-of-life care. Approximately 4,400 physicians are currently board certified or members of the American Academy of Hospice and Palliative Medicine, but it is estimated that an additional 6,000–18,000 hospice and palliative medicine physicians are needed to staff the current number of hospice- and hospital-based palliative care programs at appropriate levels (Lupu 2010). Americans spend a great deal of money on end-of-life care: nearly, 6% of Medicare patients who die each year make up 27–30% of Medicare costs (Emanuel 2013), but increased spending does not guarantee they experience the kinds of deaths they might prefer. Most Americans (75%) wish to die at home but only 20% do, with the majority dying in institutions after an explicit decision is made to limit care.[1] Only 20–30% of Americans report having an advance directive, such as a living will, that specifies their end-of-life care preferences. Even worse, only 25% of physicians in a recent study knew that their patients had an advance directive on file (Tillyard 2007). While more than 80% of patients want to avoid hospitalizations and high-intensity care at the ends of their lives, their wishes are often overridden by patient-designated surrogates and next-of-kin proxies, who incorrectly predict the patient's end-of-life treatment preferences (Shalowitz et al. 2006), and ironically even by doctors who would choose a "no-code" status for themselves but tend to pursue aggressive, life-prolonging treatment for their patients (Periyakoil et al. 2014).

Our goal in this book is to enrich the practicality and nuance of ethical analysis applied to the moral problems surfacing in contemporary end-of-life care. Each case presents a unique and ethically problematic situation in which medical care decisions at the end of life defied easy, neat, or universal resolution. While some of the lessons to be learned are generalizable, each case also reveals issues that reflect particular configurations of patient characteristics, organizational structures, political climates, medical cultures, and interpersonal relationships. There are no easy solutions or ones that will be satisfactory to all stakeholders. Each of the cases presented involved real people, with varying intentions, trying to make decisions they could live with, even after the patient died and the headlines faded. These cases provide lessons in how ethical principles, precedents, and virtues must also accommodate relationships, family dynamics, political realities, and social conventions. We believe that their casebook offers several unique things: (1) specific, real-life cases not made available heretofore; and (2) a wide-angle view of the apparent problems or issues at hand in each case, beholden to no particular "school" or ethical approach, yet insistent upon thorough and rigorous argumentation in developing the best analysis, approach, or resolution possible.

We have both worked in the healthcare arena for more than 25 years and have been active participants in efforts to improve end-of-life care. Lori A. Roscoe is Associate Professor in Health Communication in the Department of Communication at the University of South Florida (USF) where she also earned her Ph.D. in Aging Studies. Her dissertation research examined the ethical, clinical, and psychological factors that influenced the clients of Dr. Jack Kevorkian to request his assistance in ending their lives. Her current research focuses on the communication issues that complicate end-of-life decisions. Dr. Roscoe teaches undergraduate and graduate classes in communication ethics, health communication, aging, and end of life; she spent 5 years in the Office of Curriculum and Medical Education at the USF Morsani College of Medicine developing and implementing classes in medical ethics and humanities, geriatrics, and professionalism. Dr. Roscoe is also on the Executive Committee of the Center for Hospice, Palliative Care, and End-of-Life Studies at USF, a university–community partnership research center that funds pilot grants, provides assistantships for doctoral students conducting end-of-life research, and has sponsored national conferences on end-of-life issues including a physician board review course for certification in hospice and palliative medicine.

David P. Schenck is an Emeritus Professor of USF, who earned his Ph.D. in French language and literature from the Pennsylvania State University. He followed the usual academic career path of teaching and research, responding along the way to interesting challenges that resulted in 15 years of service devoted to college and central administration. At the same time, he developed a keen interest in bioethics, pursuing this field over many years at Georgetown University, and spending most of the last decade of his career teaching biomedical ethics in both the Honors College and Department of Religious Studies at USF. Since 2000, he has also held an affiliate appointment in the USF Department of Otolaryngology as its ethicist; his research in the Head and Neck Surgery Program of that department has focused primarily on oral cancer in Hispanic migrant farmworkers.

We have each served on various hospital and hospice ethics committees over the past 25 years, both continuing service on one or more today. We are both also intimately familiar with Institutional Review Boards (IRB). Dr. Roscoe served as a member of an IRB of a large, state-supported, comprehensive, graduate, and research institution which counted, within its broad spectrum of academic divisions, colleges of medicine, nursing, and public health, as well as numerous institutes and research centers dedicated to special areas of focus in health care. Dr. Schenck is currently a member of the IRB of a large urban, private, not-for-profit, local multi-hospital system that reviews and maintains oversight of hundreds of new and ongoing funded and unfunded clinical studies annually.

We have worked closely together as colleagues for many years. In 1999, we found ourselves serendipitously housed in what was then known as "The Ethics Center" on the main (Tampa) campus of USF, where one interesting conversation led to another intriguing idea, and before a year had passed a new course in biomedical ethics had been approved and was being taught jointly by the two of us.

The success of this work led to further collaboration, including a jointly authored, delivered, and subsequently published conference paper; a subsequent joint publishing effort; joint attendance at workshops and seminars of mutual interest; additional shared presentations and research efforts; lengthy service together on a particular ethics committee; and collaborative work of various kinds with USF head and neck surgeons.

We have spent many, many hours discussing cases with which we have been intimately familiar, working through them again and again, reviewing "what went wrong" or "what could have been done better" or "how this kind of thing could be avoided next time." We have reviewed our collaborative work tirelessly, and have bravely asked one another for critical reviews of work to be submitted for publication or funding, begging for honest, brutal candor in response, trusting that the other will indeed be forthright. We have come not only to trust one another but also to believe that we are very much in the same mode of thinking with regard to human values, the fundamental principles that should guide biomedical ethical decision-making, and the role of the virtues in this process. We share very similar views on such important issues as to what it means to be sick and to suffer; the significance behind the terms curing, healing, and wholeness; the roles of both patient and physician; and an understanding of the goals of medicine. Yet, differences in perspectives, experiences, and training have emerged as well. Bioethical concerns often give way to communication difficulties. Lack of effective communication between doctors, patients, and family members creates untenable situations for all concerned. Conversely, sometimes what appears to be competent communication between doctors and patients can mask important underlying ethical problems.

We believe that ethical dilemmas, especially those found in complex or complicated cases, may involve both bioethical issues as well as those in communication. What can sometimes appear to be an ethics case may in fact be most readily resolved by focusing on communication between the various parties such as patients, family members, physicians, nurses, and other members of the healthcare team. Yet, other cases that appear to contain intractable problems in communication might be resolved when all parties concerned focus on the ethics involved—how best to honor a loved one's treatment preferences, for example. The principles and practices of bioethics, coupled with case analysis, allow us to focus on ways to honor patient autonomy, examine the balance between beneficence and non-maleficence, the virtues expected in the helping professions, conflicts between ethics and law, and the increasing need to focus on just resource allocation, health disparities and access to healthcare resources. Communication theory is particularly attuned to contextual features, systems, and relationships, which simultaneously complicate already complex patient care situations, and which may also provide the resources needed for a satisfactory resolution. This book brings bioethical principles, concepts, and reasoning into conversation with communication concepts such as social construction, sense-making, framing, and relational dialectics. This framework thereby provides readers with opportunities to fit themselves into the situations described in order to cultivate unique and divergent explanations that reflect the complex realities of contemporary medical practice, including the changing relationships between patients and practitioners, shifting perceptions of the role of technology in human existence, and evolving social ideals about life and death.

The cases chosen for inclusion in this book are those we see as containing more than the usual complexity to be found in casebooks or journals presenting case discussions, where ethical dilemmas may often appear to be the product of conflicts between principles, values, cultural/national/racial/religious, or other significant differences between parties, or where seemingly problematic ethical issues may ultimately be rather easily resolved through better communication. Some might call these cases "wicked problems" because they each defied easy resolution. The cases here have been chosen because more than one kind of conflict appears to be in play, either ethical or communicative in nature, or both, especially where there may appear to be no satisfactory or acceptable resolution. The claims of multiple stakeholders had to be taken into account, almost always against a backdrop of intense scrutiny from the legal system, the media, as well as from religious organizations.

We have found all too often, however, not only among students but also among seasoned professionals, a tendency to look for answers that offer the "easy out" solution, particularly if one can point to a principle, a rule, or even a law that would seem to overrule virtually everything else, but which in fact has only the effect of so oversimplifying things that valuable nuances are lost and critical issues of human concern are not fully explored. The purpose of this book, then, is to provide opportunities for careful examination of complex cases through a broad, somewhat hybrid, approach that does not promote or embrace any particular stance, "school," or heretofore identified critical perspective on ethics, such as principle ethics, virtue ethics, feminist ethics, or casuistry, to name a few. We do not, therefore, attempt to follow any particularly prescriptive approach to case analysis, and we certainly do not attempt to create our own method. We describe our methodology as one that is grounded essentially in the historical growth and development of principle ethics, coupled with narrative ethics and its emphasis on ethical reasoning derived from stories, in an environment that encourages "outside the box" approaches to problem-solving. Narrative ethics (Geisler 2006) is an umbrella term for ethical reasoning derived from stories, whereas principlism (Beauchamp and Childress 2012) employs and balances abstract principles such as autonomy, beneficence, non-maleficence, and justice to determine right action. Each approach makes a unique contribution to understanding moral life and the process of ethical decision-making in healthcare situations. As McCarthy succinctly states, "a good principlist has narrativist tendencies and a good narrativist is inclined toward principlism" (McCarthy 2003).

Principlism is a more traditional approach to ethical reasoning that generally takes respect for persons (or autonomy) as the principle of most importance in bioethical decision-making. The focus is on understanding and putting into practice what the patient would want for him- or herself. Other principles are brought to bear as well. Beneficence challenges us to do good for others, while non-maleficence reminds us to "first do no harm." The principle of justice demands that we treat everyone equally and remain mindful of issues of cost and resource allocation.

Narrative ethics regards moral values as an integral part of stories and story-telling because narratives themselves implicitly or explicitly ask the question, "how should one think, judge, and act—as author, narrator, character, or audience—for the greater good." Our approach focuses primarily on the ethics of "the told" where we are most concerned with exploring the ethical dimensions of characters' actions, especially in the conflicts they faced and the choices they made, and how the narrative's plot unfolded to reveal the ethical issues faced by all individuals involved.

The cases here allow readers to practice ethics by applying both ethical prin-ciples and narrative competencies to understand, interpret, and determine right actions. The case narratives are written to expose the perspectives of multiple stakeholders and to demonstrate that medical plotlines in many end-of-life situa-tions are far from predictable, controllable, or generalizable. The literary skills that allow readers to understand and interpret stories also help reveal the ethical issues embedded in a case narrative, while the ethical principle of autonomy helps center the questions to be resolved within the context of a patient's individual beliefs, culture, and life events. The combined approach also allows contextual features of each case, such as family dynamics, the political scene, the conflicting priorities of various professional interests (medical specialists, nurses, social workers, and hospital administrators), and institutional cultures and practices to be taken into account. What follows are all actual cases. Permission to use them has been granted either by the institutions, or officials, or physicians involved, provided all identifiers be stripped, and every effort has been made to ensure that. In some cases, not only have names or initials been altered but dates and locations have been changed as well. The use of initials may seem to be an impersonal choice, but this was done deliberately. Names convey a great deal of information—social class, age, and ethnicity, among other characteristics. Whatever we knew about the patient is disclosed in the discussion, if not in the case description, but it can be instructive to deliberate about the ethical dimensions of a difficult case, free from attributions about certain kinds of patients. Identifying patients and other individuals by initials hopefully encourages readers to analyze the kind of inferences one may make about certain kinds of patients in certain kinds of situations.

Each case is described in some detail, but that should not be understood to mean that it would necessarily include everything a reader might wish to know, as may often be the case with real-life ethical dilemmas. Case descriptions are followed by discussion questions designed to help focus a conversation about the issues pre-sented in the case. Following the discussion questions, we provide responses from their respective disciplines, Dr. Schenck from Bioethics, and Dr. Roscoe from Health Communication.

These cases are not examples of the application of specific ethical principles, the dilemmas that attend to particular kinds of patients, or situations that involve only certain healthcare professionals. They all involve hospitalized patients at the ends of their lives, and they are illustrations of the complexities of human decision-making and ethical justification against a backdrop of real-life circumstances: the realities of media coverage, politics, a heterogeneous public, and the lack of civil dialogue in almost every avenue of public life. Our hope is that the sharing of these cases may help those in the trenches of health care (or those about to be) to learn about how good people, in difficult circumstances, can strive to reach ethical, legal, and medically appropriate solutions when confronted with a vast and difficult range of circumstances. None of them provides easy answers, but they should allow readers of all kinds to test their ethical judgments, moral commitments, and knowledge of the law and professional codes of conduct against the real dramas that play out every day as patients and families, physicians, and other healthcare providers attempt to navigate the rocky and difficult terrain of end-of-life care. Many of these cases are cautionary tales, and we hope that sharing them will allow better solutions to be imagined and enacted.

Notes

[1]For more information, see http://www.pbs.org/wgbh/pages/frontline/facing-death/facts-and-figures/.

Tampa, USA   Lori A. Roscoe
                                                                                   David P. Schenck

References

Beauchamp, Tom L., and James F. Childress. 2012. *Principles of biomedical ethics* (8th edition). New York: Oxford University Press.

Caplan, Arthur, James. J. McCartney, and Domenic A. Sisti. 2006. *The case of Terri Schiavo: Ethics at the end of life*. Amherst, NY: Prometheus Books.

Cruzan v. Director, Missouri Department of Health, 497 U.S. 261 (1990).

Emanuel, Ezekiel J. 2013. Better, if not cheaper, care. *The New York Times*. http://opinionator.blogs.nytimes.com/2013/01/03/better-if-not-cheaper-care/

In re Quinlan (70 N.J. 10, 355 A.2d 647 (NJ 1976).

Gage, Barbara, Susan C. Miller, Kristen Coppola, Jennie Harvell, Linda Laliberte, Vincent Mor, and Joan Teno. 2000. Important questions for hospice in the next century. U.S. Department of Health and Human Services, March 2000. http://aspe.hhs.gov/daltcp/reports/impques.pdf

Geisler, Sheryl L. 2006. The value of narrative ethics to medicine. *The Journal of Physician Assistant Education* 17: 54–57.

Lupu, Dale. American Academy of Hospice and Palliative Medicine Workforce Task Force. 2010. Estimate of current hospice and palliative medicine physician workforce shortage. *Journal of Pain and Symptom Management* 40: 899–911. doi:https://doi.org/10.1016/j.jpainsymman.2010.07.004.

McCarthy, Joan. 2003. Principlism or narrative ethics: Must we choose between them? *Journal of Medical Humanities* 29: 65–71.

NHPCO's Facts and Figures. 2015. The National Hospice and Palliative Care Association. http://www.nhpco.org/hospice-statistics-research-press-room/facts-hospice-and-palliative-care

Oregon Public Health Division. 2015. http://public.health.oregon.gov/ProviderPartnerResources/EvaluationResearch/DeathwithDignityAct/Documents/year17.pdf

Periyakoil, Vyjeyanthi S., Eric Neri, Ann Fong, and Helena Kraemer. 2014. Do unto others: Doctors' personal end-of-life resuscitation preferences and their attitudes toward advance directives. *PLOS ONE.* doi:https://doi.org/10.1371/journal.pone.0098246

Roscoe, Lori A., Julie E. Malphurs, L. J. Dragovic, and Donna Cohen. 2000. Dr. Kevorkian and euthanasia cases in Oakland County, Michigan, 1990–1998. *New England Journal of Medicine* 34: 1735–1736.

Shalowitz, David I., Elizabeth Garrett-Mayer, and David Wendler. 2006. The accuracy of surrogate decision makers: a systematic review. *Archives of Internal Medicine* 166: 493–497.

Tillyard, Andrew R. J. 2007. Ethics review: 'Living wills' and intensive care—an overview of the American experience. *Critical Care* 11: 219. doi:https://doi.org/10.1186/cc5945

Vacco v. Quill, 521 U.S. 793 (1997).

Washington v. Glucksberg, 521 U.S. 702 (1997).

Acknowledgements

I acknowledge the never-ending love and support of my husband, Eric Eisenberg; my sons, Evan and Joel Eisenberg; my parents, Lucille and Charles Roscoe; my sister, Nancy Lloyd; my sister-in-law, Danya Lane; and my aunt, Florence Millon. All of you bring great joy to my life. Thank you for believing in me and listening to my stories. And to my grandmother, Angeline Florkowski, who always encouraged my love of writing.

Tampa, Florida   Lori A. Roscoe, Ph.D.
September 2017

I acknowledge here two very special people for their unfailing support of my work. The first is my dear wife, Mary Jane, whose enduring patience was tested beyond reason, especially at a time in her life when she needed my undivided attention more than ever; no man ever deserved what this loving woman gave. The second is the late Dr. Edmund Pellegrino, my friend and mentor in biomedical ethics, who nurtured and encouraged me in this field, and who taught me what it means to be sick and to suffer, and what it genuinely means to care for our fellow sufferers.

Tampa, Florida   David P. Schenck, Ph.D.
September 2017

Recommended References for Beginners

This casebook is composed of complicated current cases, whose analysis presupposes a basic or intermediate level of familiarity with ethical principles, case-based reasoning, and foundational cases that created the current underpinning on which to base ethical judgment. Where possible, we have included references and summaries that will assist readers in understanding the cases presented here; in addition, we include a partial list of other ethics casebooks and other volumes that provide an excellent background for interested readers.

- Beauchamp, T. L., & Childress, J. F. (2012). *Principles of biomedical ethics* (7th ed). New York: Oxford University Press.

Many bioethicists consider this to be the foundational text for understanding the history and current trends in bioethical reasoning. This edition describes in detail a principlist approach to bioethical reasoning, which we adapt in our case analysis, and which will be familiar to those who practice medicine, teach biomedical ethics in professional schools, or serve on ethics committees.

- Pence. G. E. (2004). *Classic cases in medical ethics: Accounts of cases that have shaped medical ethics, with philosophical, legal, and historical backgrounds* (4th edition). New York: McGraw-Hill.

This book is written by a professor of philosophy and begins with an overview of moral reasoning and ethical theories in medical ethics. This is an excellent casebook to familiarize students with the classic cases in a variety of areas, and the cases include details about the court cases that helped resolve these cases. Classic cases are important to understand but they do not always prepare interested persons to anticipate the outcome of more current cases in which technology, law, and medicine, among other important contextual features, have changed since these classic cases were resolved.

- Ford, P. J., & Dudzinski, D. M. (2008). *Complex ethics consultations: Cases that haunt us*. New York: Cambridge University Press.

This casebook is an edited collection of 28 cases submitted by authors representing diverse disciplinary viewpoints. Each chapter includes a case narrative, professional reflections, haunting aspects, outcomes, discussion questions, and references. All cases are presented from the point of view of the ethics case consultant who was involved with the case and who authored the case for inclusion in this collection.

- Fry, S. R., Veatch, R. M., & Taylor, C. R. (2011). *Case studies in nursing ethics* (4th edition). Burlington, Massachusetts: Jones and Bartlett Learning.

This casebook is designed specifically for students in upper level undergraduate and graduate-level nursing courses, and presents basic ethical principles and specific guidance for applying these principles in nursing practice. Each of the 150 case studies allows readers to develop their own approaches to the resolution of ethical conflict and to reflect on how the traditions of ethical thought and professional guidelines apply to the situation. This is a good resource for nurses but is not specifically focused on end-of-life patient situations.

- Gervais, K. G., Priester, R., Vawter, D. E., Otte, K. K., & Solberg, M. M. (Eds). (1999). *Ethical challenges in managed care: A casebook.* Washington, D. C.: Georgetown University Press.
  ISBN-13: 978-0878407194
  ISBN-10: 0878407197

This book contains 20 case studies that present a wide range of ethical challenges that explore the goals, methods, and practices of managed care. Accompanying each case are questions for consideration and a pair of commentaries by prominent contributors from diverse fields. Through the cases and commentaries, this book clarifies the internal workings of managed care, explains relevant concepts, and offers practical, constructive guidance in addressing ethical and policy issues. It is designed primarily for those managing the delivery and financing of health care and for students and practitioners of the health professions and health administration.

- Thomas, J. E., & Waluchow, W. J. (1998). *Well and good: A case study approach to biomedical ethics* (3rd edition). Peterborough, Ontario: Broadview Press.

This book includes both real-life cases as well as classic cases, most of which involve nurses and other allied health professionals. The cases in the main body of the book are accompanied by the editors' discussions of the issues involved.

- Veatch, R. M., & Flack, H. E. (1997). *Case studies in allied health ethics.* Upper Saddle River, New Jersey: Prentice Hall.

This book contains case studies based on the actual experiences of practicing allied health professionals in various fields (such as dietetics and occupational therapy). The book is somewhat dated and does not always reflect changes in these fields, and it can be helpful to students and practitioners in allied health fields.

For readers interested in more background on family dynamics and family communication, the following are suggested references:

- Hoffman, L. (1981). *Foundations of family therapy: A conceptual framework for systems change.* New York: Basic Books.

Hoffman's classic book provides a synthesis of themes and concepts around which family theory and therapy have evolved. Starting with Gregory Bateson's ideas on social fields, the book examines concepts that have come to family therapy from general systems theory. The book also explores the major schools of family therapy.

- Vangelisti, A. L. (ed). (2003). *The Routledge handbook of family communication*. New York: Routledge.

The Routledge Handbook of Family Communication offers a comprehensive exploration and discussion of current research and theory on family interaction. It integrates varying perspectives and issues addressed by family researchers, theorists, and practitioners, and offers a unique and timely view of family interaction and family relationships. Research of issues key to understanding family interaction is synthesized, and the various theoretical and methodological choices made by researchers studying family communication are analyzed. The handbook highlights the work of scholars across disciplines—communication, social psychology, clinical psychology, sociology, family studies, and others—to capture the breadth and depth of research on family communication and family relationships.

Contents

Part I
Beginnings/Endings: Complex Issues with Pregnancy, Newborns, and Young Children

End-of-life cases involving pregnancy, newborns, and young children are difficult. There are at least two patients to consider—at a minimum, a mother and child. Most often, there are other family members to consider. Birth is usually a joyous event in a hospital, but not always. The cases discussed here remind us that sometimes hospitals are sites of celebration, but more often sites of illness and dying. Death is nearly always hard to accept, but when a child is dying, the emotions, medical complexities, communication challenges, and ethical dilemmas can multiply.

The cases discussed in this section involve issues such as the biases healthcare professionals may hold about families with different lifestyles, values, or cultural traditions; appropriate decision-making roles for parents who are too young to legally make medical decisions for themselves; parents who have unreasonable expectations about what medical care might achieve for their child; and the ethics surrounding postmortem sperm retrieval. A short summary of each of the five cases in this section follows:

Case 1—Does the Nearness of Death Diminish the Value of a Life?

A sex worker who was pregnant with her fourth child was diagnosed with a fatal tongue tumor. Balancing her treatment preferences along with the needs of her unborn child was further complicated by difficult family relationships and a history of poor access to healthcare resources.

Case 2—When Cultures Collide and a Newborn Almost Dies

A graduate student studying in the U.S. from China gave birth to a baby girl with Down syndrome and an esophageal fistula. The mother refused to consent for surgery for the baby and explained that her disabled daughter would not be welcomed by extended family members when she and her husband completed their degrees and returned to China.

Case 3—When the Family Won't Decide

A teenage mother gave birth to a baby with multiple congenital abnormalities including alobar holoprosencephaly (where the brain does not develop into left and right lobes), cleft lip/palate, severe gastroesophageal reflux, failure to thrive, and seizure disorder. The baby's young parents each lived with their own parents and refused to make medical decisions or communicate with the medical team.

Case 4—Aggressive Treatment for a Child's Inoperable Tumor

A 4-year-old boy was diagnosed with an anaplastic Wilms tumor, inoperable because of its massive size and proximity to vital organs. The parents insisted on continued aggressive treatment, despite its high risks, harsh side effects, and limited efficacy. The parents' religious beliefs—one is a Jehovah's Witness and the other relied only on nonmedical approaches to health care—further complicated treatment decision-making.

Case 5—Is There Life After Death? A Case of Postmortem Sperm Retrieval

A young widow asked her late husband's physicians to retrieve his sperm posthumously so she could conceive a child in his memory, and then changed her mind when she fell in love with another man.

Case 1—Does the Nearness of Death Diminish the Value of a Life?

1

I had been standing just behind Dr. C, a head and neck surgeon, watching him perform a laryngectomy when he got a call from his chief resident at City General Hospital about a seriously ill patient who had recently been admitted. When he had concluded his brief conversation on speaker phone, Dr. C suggested I would find CS a "very interesting case" and that I should go to City General and visit her. But upon hearing the summary of her medical and social history, I said, "Well, she's never going to talk to me about those things!" "Yeah, she will," Dr. C countered. "She's very open about her life. Go see her." I did the next afternoon.

I knocked gently on her door and got back a barely audible response. I peeked in to find a huddled form curled into a fetal position. When I apologized at disturbing an obvious nap and said I'd come back later, the patient raised her head to object that she wasn't really asleep, and asked who I was. I identified myself as Dr. X, the bioethicist with the head and neck program, a colleague of her surgeon's, that I had merely come to look in on her to see how she was doing, but that I didn't wish to disturb her and could certainly come back another time. She insisted that she would like some company, but asked if I could get her nurse to give her some Lidocaine rinse to numb the pain in her tongue first so that she would be able to talk more comfortably. With that quickly accomplished she became quite chatty, and I soon discovered what Dr. C had meant. CS was a very interesting person indeed.

She was 37 years old and had presented with a large mass in the right oral tongue and enlarged nodes in the right neck. She was extremely uncomfortable, requiring high doses of pain medication. A biopsy had come back as positive for squamous cell carcinoma. She identified herself as a prostitute with a 20-year history of drug use. She had recently been released from a one-year jail term for possession of cocaine. She denied use of alcohol. She had a 20-year history of tobacco use. She was 17 weeks pregnant at the time I visited her in the hospital.

CS had three living children, none by the same father: a girl of 12 living with her (CS's) mother; a 9-year-old son living with her sister; and a 3-year-old daughter living in foster care. CS stated that her mother had recently lost both her own job and house, so that her mother and 12-year-old daughter were now living

© Springer International Publishing AG 2017

L. A. Roscoe and D. P. Schenck, *Communication and Bioethics at the End of Life*, https://doi.org/10.1007/978-3-319-70920-8_1

temporarily in the sister's house. The sister, however, had said that her mother and niece would have to leave, that she'd had enough of "this dysfunctional family." She had said, nonetheless, that she would continue to care for CS's son. From what I could glean from CS's tale, she had never taken full responsibility for any of her three living children and must have lived alone or at someone else's expense her entire adult life.

Despite her past and present difficulties, CS was excited about her pregnancy, and she was eagerly collecting new baby clothes, all neatly laid out on a table beside her bed. The baby's father had paid her no visits since her admission to the hospital, yet CS seemed positive about the future. She told me about how she had most recently been living in a crack house, of her considerable anxiety over this, and of her eagerness to move elsewhere. She said she had another male friend who was a "good man," not the baby's father, and that he had said he would provide a home for the three of them once the baby arrived. This seemed to give her hope.

But nothing quite seemed to fit into the picture I was developing here. CS probably had no more than a high school education, if that, yet she appeared very intelligent, was highly articulate and well-spoken, and bore absolutely no traces of the coarseness or street language one might have expected of someone with her background and who had been largely left to her own devices as a latch-key child whose mother worked, whose father had abandoned the family, and who had witnessed other men beat her mother. "I don't blame anyone but myself," she told me, "I brought this all on myself by making bad choices." Still, she was convinced that "God was going to heal her" of her cancer and, in fact, asked me to pray for her as I left her room that day.

Two days later Dr. C performed a partial glossectomy (surgical removal of part of the tongue) and a modified radical neck dissection, as well as a gastrostomy to provide artificial nutrition and hydration during CS's recovery. This relieved her of the immediate pain caused by the tongue tumor and enlarged neck nodes. Her spirits were high post-operatively, and she did well at first. She was discharged from the hospital after four days; her pregnancy was then estimated to be about to enter its 19th week. CS's mother had apparently managed to find a new apartment quickly and agreed to take her daughter in temporarily while she convalesced. CS was scheduled to begin routine radiation therapy six weeks later when her pregnancy would have been at about 25 weeks, but as that date approached she suddenly developed new nodes in the operated neck. It was determined that no treatment could be of any curative value at that time, that palliative measures alone were called for, and the scheduled radiation protocol was canceled.

When she returned to the clinic for the six-week post-op visit, her mother and 12 year-old daughter both accompanied her. On that visit she no longer claimed that God was going to heal her, and she expressed a real fear of dying, at one point breaking into tears. Her daughter then movingly and convincingly became mother to the child, taking her mother's hand, looking her directly in the eye, calling her mother by her first name, and saying with all the confidence of someone twenty years older, "It's OK, C_____, everything's going to be OK." It was clear to her physicians, however, that she was obviously dying of her disease.

By this time CS required increasing amounts of narcotics to control her pain, but she could not be relied upon to follow the prescribed dosages and would use up her supply before time for renewal. It was unclear where she was living at this point, but perhaps her mother had again taken her in on a temporary basis. She did begin radiation therapy for palliation, but after several weeks was taken into the hospital in a rapidly deteriorating condition. The severity of her pain required that she be sedated, but this and the use of other medications had to be titrated carefully so as to minimize effects on the baby.

At this point CS's mother asked that all efforts to keep her daughter alive be discontinued. "My daughter made it clear to me several weeks ago that if she ever ended up in a condition where she was clearly dying, she wanted to have all life support stopped!" she said. Inasmuch as pregnancy now made itself the primary medical issue in this case, an obstetrician had become the attending physician. He categorically refused the mother's request for withdrawal of all life support stating that he had "an obligation to save the baby." The mother asserted, nonetheless, that her daughter's wishes were not being respected, and that in any case, "I will not care for 'another crack baby' as I have done twice before. The new baby will have to go into foster care with the three-year-old." CS's mother thus effectively walked away from her daughter's situation.

CS was then at approximately 30 weeks of pregnancy, and it was at that point that I saw her for the last time. The afternoon I visited the ICU a technician was monitoring the baby who appeared to be doing fine. The sedation being administered to CS for pain rendered her unable to communicate, but her movements suggested sedation might not have been sufficient to relieve all discomfort. Several days later the baby showed signs of distress. A C-section was performed, and a viable infant was delivered and placed in the NICU. All life support for CS was then discontinued, and she expired within 48 h. The infant remained in the NICU for 3 months before being discharged to foster care.

Discussion Questions:

1. Is it possible for health care professionals to put aside stereotypes and biases and treat patients without these influences?
2. Is it possible to fully acknowledge the autonomy of a pregnant woman whose end-of-life treatment preferences might compromise optimal conditions for her fetus?
3. What is the appropriate relationship between law and ethics? Must ethical judgment always follow the law? Should law follow ethical principles?

A Bioethicist Responds

One might at first be tempted to view this case as one of those "no-brainers." After all, what is there really to discuss here? This is an unfortunate situation where a young woman, by her own admission, has made terribly poor choices her entire life such that not only she, but now four children, her family, and even society must pay a price of some magnitude, and there is suffering all around. What could possibly be done to avoid any of it, especially at the bitter end, other than to grant CS's mother's request for withdrawal of all life support prior to delivery of the baby? The mother had stated that such would have been in keeping with CS's autonomous request, but there is clear state law prohibiting a proxy acting in such a manner for an incapacitated pregnant patient.[1] Thus, there would seem to be little left to say. We might look to a recent case in Texas, where state law required that Marlise Munoz, a brain-dead pregnant woman, be maintained on life support until such time as her fetus could be safely delivered (Fernandez and Eckholm 2014a, b). Both Munoz's parents and husband believed that this was not only against her treatment preferences, but also a ghastly spectacle that only served to prolong and publicize their private tragedy. Ms. Munoz, 33, was only 14 weeks pregnant when she arrived at the hospital. After her 22nd week of pregnancy, doctors found that the fetus was not viable, which the hospital acknowledged in court documents. The fetus suffered from hydrocephalus—an abnormal accumulation of fluid in the cavities of the brain—as well as a possible heart defect, and the lower extremities were deformed. With sufficient time and publicity, the medical facts prevailed: A brain-dead woman was not a suitable incubator for a developing fetus, and once this was established medically, life support was allowed to be discontinued.[2]

Yet if law, albeit one legitimately established for good reasons, becomes the deciding vote in difficult cases, we may deny ourselves the benefit of careful reflection that could prove beneficial in thinking through other thorny issues. This is not to suggest one fault the obstetrician, and propose instead that he should have granted CS's mother's request despite his resulting commitment of an obvious crime, but it is to serve as a reminder that law ideally *follows* ethics, not the reverse, and that this case should not, therefore, be summarily dismissed without further discussion. Following the letter of the law in this case does not automatically mean that ethics has been given its due.

The case is noteworthy in the first instance because of the very character of the patient. The narrator in the case had been primed by the head and neck attending physician to expect to see a particular "type" of 37 year-old woman, a prostitute, who had already given birth to three children, who was now pregnant again and in her 17th week, and who was also an admitted 20-year drug user. He is likely to form a certain image, no doubt taken from TV and movies, of what this type of woman might look like, how she might behave, how much education she might have, what level of society she might come from, what her values might be, how she talks, and what kind of language she might use. But what he found was a person very different from the one he had anticipated: He finds an intelligent, articulate individual who, with surprising insight, it appears, blamed neither other people nor society for her plight but rather held herself accountable for making bad choices. Moreover, she seemed to believe that God would heal her, and she asked for prayer on her behalf. This may well be "fox-hole religion," or nothing more than denial, but one would have to be severely myopic were one not to see that CS was clearly a victim crying out for help now that she was in the hospital, in considerable pain, and about to undergo very serious surgery.

Fortunately for CS, she happened upon a highly skilled and very kind, caring head and neck surgeon who would do whatever it took to save her life or to give her the best quality of life possible for what may remain of it. Yet he was only part of the team that must care for her. People like CS, with their lifestyles and histories of bad choices, especially when those choices affect others such as their own infants, are not always viewed as the most desirable patients. This is especially true when they are unfunded, as was CS. Whether we wish to admit it or not, and regardless of how well-intentioned we would hope for all those interacting with CS to be, discrimination in some form will inevitably enter this picture. This could be anything from indifferent or rude treatment by clerical personnel in business offices or clinics, to delayed attention or nonchalant attitudes regarding pain or symptom relief on the part of physicians. Add to that her long-standing addiction, thus making physicians wary of her reliability in following instructions for taking narcotic pain medications, her almost predictable early requests for refills, only to the chief resident's frustration, and she soon becomes *persona non grata*.

Another ethical issue this case presents is how we might respect the autonomy of a terminally ill pregnant woman, while still adhering to the legal (and ethical) requirement to protect the health and life of the fetus. The head and neck team demonstrated exemplary compassion, and offered her the best care and treatment a teaching hospital in a major metropolitan medical center could, with fairness and justice despite the fact that she had virtually no resources whatsoever, knowing that their story together would come to its natural conclusion when her disease eventually escaped their control and raced ever faster toward its own predictable end. The team was hardly unaware that there were, in fact, two patients in this case, but although they must do no harm to the one as yet unseen, their primary concern was for the one they had been charged to see and had come to know.

Yet when it was clear that the head and neck surgeons had done all they possibly could for CS and handed her off to obstetrics, everything about this case changed. The obstetrician's central concern naturally became delivering the healthiest infant possible. With CS sedated and very close to death at that point, she inevitably became a non-person in the eyes of the obstetrical team. This was not the result of coldness or lack of empathy, but rather an example of what happens when medical care is segmented by body part, diagnosis, and medical specialty. Everyone involved in CS's case looked after their own "part" and no one was charged with oversight of the whole situation. What a blessing, one might say, for there was nothing to be done at this point in any case about this terribly unfortunate individual and her tragic life; at least one could focus on rescuing a potential new life. And yet it is precisely here where a very significant ethical issue comes into focus, namely that even a terminally ill pregnant woman may not be treated as merely the means to an end. This was not just the end of a bad situation, the blessed relief, and the resolution to everything. This was in fact an epiphany, the realization that everyone had failed CS. One might do well, therefore, to ask *how, when,* and *where* CS could have been provided assistance that would perhaps have made her dying process better.

The complexity of CS's case, with its attendant frustrations, might in fact have been seen as the source of some resolution to what was viewed, most certainly by

CS herself, as her own tragic end. As mentioned in the case report above, she had been excited about this pregnancy, despite the fact that this would be her fourth baby without a stable home life, and despite the fact that the father had not once visited her in the hospital. CS also told me she had a male friend who had promised he was going to find an apartment for the three of them once the baby arrived. What this last statement of hers, coupled with her expression of joy at the impending arrival along with her new collection of baby clothes really told me fundamentally, however, was that she embraced a strong feeling of hope. And when I saw her in the clinic at the six-week post-op visit, with a huge nodal recurrence in her neck, her tears and the fear in her voice, as she said to me, pleadingly, "I don't want to die!" reinforced the notion that she clung to some hope her life could be saved. In retrospect, I think we all missed an opportunity for healing for this patient.

CS had earlier expressed a faith in God. Foxhole religion? Perhaps, but whether she had a genuine faith or not, it would have been useful for a skilled chaplain or social worker at that point to begin working with this dying patient to assist her with the process of writing the remaining chapters of her life story, how she would like them to look, visioning them in a way that would be satisfactory both for her and for her new baby, both with and without one another, so that she did not have to simply be sedated without any benefit of closure. Had there at least been some attempt at something such as this, the case of CS would not have to be seen as just one more tragic case involving a person who is in great measure clearly responsible for her own situation. It could be seen instead as a situation with sufficient texture to provide the narrative threads with which CS just might be able to come to terms with her situation realistically and tell how she could see the end of her life and the beginning of her baby's life in a way satisfying to her. And, in the best of all possible worlds, she might just be able to experience something of the distinction between curing and healing, all the while being healed as she died.

A Health Communication Scholar Responds

The concept of framing, described in detail by sociologist Erving Goffman and employed throughout communication scholarship, suggests that how something (or someone) is presented to the audience (called "the frame") influences the choices people make about how to process that information (Goffman 1974). Goffman suggested that people interpret what is going on through a primary framework which is taken for granted by the user. Primary frameworks can be further described as either *natural*, which identify events as physical occurrences taking place without social forces, or *social*, which view events as socially driven occurrences influenced by the goals and manipulations of other social players. Frames are abstractions that work to organize or structure the meaning one ascribes to the information that has been presented. The frameworks we invoke profoundly affect how information is interpreted, processed, and communicated. Framing sets boundaries around the topic at hand, and once a frame is established, it can be exceedingly difficult to pay attention to what may lie outside the frame. Once meaning is ascribed to the information inside the frame it can be nearly impossible to see things any other way.

CS was poor and did not have health insurance. She had no stable living arrangements or a reliable partner, despite having three dependent children and a fourth on the way. Her mother and sister were tired of dealing with her and her problems, which included an occupational history as a prostitute, and drug addiction. These descriptors generally invoke a social frame that might let us see CS as a person whose own bad choices led to her present circumstances, an object lesson of how not to pursue adulthood. Indeed, her own mother suggests that such an interpretation would be appropriate. Her family's frustrations notwithstanding, however, the extent to which one's actions and past decisions are entirely under one's control is controversial. Can we control our own destinies, or are we the hapless products of our genes and our environment? Our puritanical American frameworks sometimes lead us to judgments that are less than charitable, and our culture at the present moment tolerates blaming a person's present circumstances on their inability to exercise sufficient self-control or good-enough judgment. Social policies such as welfare reform, loosening of affirmative action plans, and efforts to dismantle the Affordable Care Act, while outside the scope of this case, are evidence of our willingness as a society to let the chips fall where they may for the less fortunate members of society. Relying on this social framework may let us off the hook of being troubled by CS's fate, and allow us to be comfortable treating her as merely a convenient incubator for her unborn child without the need to respect her dignity.

The fact of CS's occupational choice to engage in sex work, if it truly was a choice, is both a reason to *not* treat her differently than those in other occupations, as well as a reason *to* treat her differently. Sex work presents a host of medical challenges, and sex workers are at risk for receiving sub-standard or no care for pregnancies and other injuries or illnesses. It is sobering to think that CS likely received better pre-natal care when pregnant with her fourth child than she had in the past because she was terminally ill. Respecting a patient's dignity means treating them with care, attention and compassion, no matter their diagnoses, past life choices, socioeconomic status, age, gender, race, religion or culture (Beach et al. 2006; Blanchard and Lurie 2004; Henry et al. 2015).

Prostitution is often exempted from the category of violence against women, but even a cursory consideration of the health consequences of prostitution demonstrates that sex work gravely impairs women's health and belongs firmly in the category of violence against women (Raymond 1999). Health consequences to women who work as prostitutes are often the same as those suffered by women subjected to other forms of violence such as broken bones, bruises, black eyes, and concussions. The sex of prostitution is physically harmful as well and can include sexually transmitted diseases (STDs) such as HIV/AIDS, chlamydia, gonorrhea, herpes, human papilloma virus and syphilis. Studies estimate that only 15% of the women working as prostitutes had never contracted one of the STDs, not including HIV/AIDS, most injurious to health (Farley 2003). General gynecological problems, including chronic pelvic pain and pelvic inflammatory disease, are also common among prostitutes.

Pregnancy is another occupational hazard. We do not know which of CS's children were planned or unplanned, conceived in romantic relationships, or were the result of professional liaisons. It is estimated that over two-thirds of the women working in the sex trades had an average of three pregnancies which they attempted to bring to term during their time in prostitution (Nutbrock 2004). Women working as prostitutes also suffer emotional health consequences including severe trauma, stress, depression, anxiety, and self-medication through alcohol and drug use. Substance abuse is common among female sex workers in large metropolitan areas; it is estimated that up to 93% of women arrested for prostitution test positive for illegal drugs (Raymond 1999). All of this is to say that CS's choice to work in the sex industry, and her willingness to accept the consequences for her actions, hide a potential multitude of factors over which she may have had little or no control. The same can be said for the social frame that allows us to see prostitutes as willing victims of their own bad choices.

CS is also viewed through the frame as a patient with advanced head and neck cancer. HPV, or human papillomavirus, appears to be responsible for the rise in cancers of the oropharynx (tonsil and base of tongue), and is related to oral sex, especially in younger patients; surely this counts as another occupational hazard of sex work.[3] Annually in the U.S., 10,000 new cases of head and neck cancer can be attributed to a certain strain of HPV. Tobacco and alcohol use are the leading causes of mouth cancers. Cigarette smoking increases the risk of head and neck cancer by 15 times compared to non-smokers, and people who use both tobacco and alcohol are at greater risk than people who use either alone. And unfortunately, 66% of the time, oral cancers will be found as late stage three and four diseases. Sex work, unplanned pregnancies, drug addiction, family issues, insufficient housing, limited access to health care—all contribute to a high level of chronic stress. Although the effect of stress on cancer growth is not fully understood, recent research has demonstrated that stress hormones, especially norepinephrine and epinephrine, can contribute to tumor progression and cancer recurrence (Palesh et al. 2007; Sood et al. 2010).

The social framework through which most patients similar to CS are viewed can make it easier to dismiss her tragic circumstances. Peering deeper into the frame as well as looking outside of it to adjust our notions of causality and choice can help restore CS's humanity (and our own). As it happens, CS was an intelligent, artic-ulate individual who in many ways resisted categorization. Perhaps even stating surprise at how unlike our stereotypical frame CS appeared is also evidence of our inability or at least difficulty in seeing past these choices to the person who was suffering their consequences, just or not. Like all of us, CS was more than the social descriptors of her situation and past life choices, or the sometimes unconscious and automatic frameworks we invoke to make sense of our social world. We must take into account CS's talk of her excitement about the pregnancy, and her perhaps unrealistic belief that she and this child will have a stable life together. Her fears and hopes deserved to be heard. Both CS and her unborn child are entitled to our compassion and care; all of our biases, fears, and beliefs in natural consequences must be put aside.

Notes

[1]At the time, the statutes of the state in which this case occurred were clear that unless the legal proxy or court-appointed guardian could demonstrate that a third-trimester abortion was necessary to protect the health or save the life of an incapacitated mother, such action was not permissible. Given that CS's mother, albeit her legal proxy, could demonstrate neither condition, and given that her insistence upon withdrawal of all life support would effectively have guaranteed the death of a viable, near-term infant, the attending had little difficulty in refusing that request.

[2]There are some important differences between the case of CS presented here, and the case of Marlise Munoz referenced in our responses. In CS's case, her fetus was viable, unlike in the Munoz case. In the Munoz case the family immediately asked for discontinuance of the ventilator, and her husband had the legal and ethical right to make that decision on his wife's behalf. In CS's case, she retained the capacity to make her own decisions nearly to the end of her life, and did want life support continued to ensure her baby's survival.

[3]See the 2015 Fact Sheet. American Academy of Otolaryngology—Head and Neck Surgery for more information (http://www.entnet.org/?q=node/1501).

References

Beach, Mary Catherine, Debra L. Roter, Nae-Yuh Wang, Patrick S. Duggan, and Lisa A. Cooper. 2006. Are physicians' attitudes of respect accurately perceived by patients and associated with more positive communication behaviors? *Patient Education and Counseling* 62: 347–354.

Blanchard, J., and N. Lurie. 2004. R-E-S-P-E-C-T: Patient reports of disrespect in the health care setting and its impact on care. *Journal of Family Practice* 53: 721–730.

Farley, Melissa. 2003. Prostitution, trafficking, and traumatic stress. *Journal of Trauma Practice, 2* (3/4).

Fernandez, Manny, and Erik Eckholm. 2014a. Pregnant and forced to stay on life support. *The New York Times*, January 7.

Fernandez, Manny, and Erik Eckholm. 2014b. Texas woman is taken off life support after order. *The New York Times*, January 26.

Goffman, Erving. 1974. *Frame analysis: An essay on the organization of experience*. Cambridge, MA: Harvard University Press.

Henry, Leslie Meltzer, Cynda Rushton, Mary Catherine Beach, and Ruth Faden. 2015. Respect and dignity: A conceptual model for patients in the intensive care unit. *Narrative Inquiry in Bioethics* 5: 5A–14A.

Nutbrock, Larry A., Andrew Rosenblum, Stephen Magura, Cherie Cillano Stephen, and Joynce Wallace. 2004. Linking female sex workers with substance abuse treatment. *Journal of Substance Abuse Treatment* 27: 233–239.

Palesh, Oxana, Lisa D. Butler, Cheryl Koopman, Janine Giese-Davis, Robert Carlson, and David Spiegel. 2007. Stress history and breast cancer recurrence. *Journal of Psychosomatic Research* 63: 233–239.

Raymond, J.G. 1999. Health effects of prostitution. Making the harm visible: Global exploitation of women and girls. http://www.uri.edu/artsci/wms/hughes/myhealt.htm.

Sood, Anil K., Guillermo N. Armaiz-Pena, Jvotsnabaran Halder, Alpa M. Nick, Rebecca L. Stone, Hu Wei, Amy R. Carroll, Whitney A. Spannuth, M.T. Deavers, Julile K. Allen, Liz Y. Han, Aparna A. Kamat, Mian M.K. Shahzad, Bradley W. McIntyre, Claudia M. Diaz-Montero, Nicholas B. Jennings, Yvonne G. Lin, William M. Merritt, K. DeGeest, Pablo E. Vivas-Mejia, Gabriel Lopez-Berestein, Michael D. Schaller, Steven W. Cole, and Susan K. Lutgender. 2010. Adrenergic modulation of focal adhesion kinase protests human ovarian cancer cells from anoikis. *Journal of Clinical Investigation* 120: 1515–1523.

Case 2—When Cultures Collide and a Newborn Almost Dies

ZQ and her husband were Chinese graduate students studying engineering at a large, urban research university. They lived quietly in an apartment close to campus, and socialized primarily with other international graduate student couples. Their plan was to move back to China upon completion of their doctoral degrees; they had already begun applying for academic appointments at Chinese universities. The husband's research was being completed at a slightly faster pace, and he had been offered an academic appointment in China, which was to begin in September. ZQ planned to stay in the U.S. and finish her own degree in time to meet him for the second academic semester the following February. Shortly after her husband successfully defended his dissertation in early spring, ZQ discovered she was unexpectedly pregnant.

The young couple discussed their options. Her husband's new job opportunity was too good to pass up. ZQ was close to finishing her degree, and she felt she would be able to continue working and living in the U.S. with the support of their graduate student friends. Her husband left in August, feeling proud that he would be able to support his wife and new baby, but sad that he would not be with ZQ during her pregnancy or their baby's birth, which was expected in December. Their friends rallied around the young couple, promising that ZQ and the new baby would be well cared for.

ZQ had health insurance through the Graduate Student Association, and she was able to receive prenatal care at the University's teaching hospital. During her first trimester, her obstetrician recommended a blood test to measure the levels of pregnancy-associated plasma protein-A (PAPP-A) and an ultrasound to measure a specific area on the back of the baby's neck, known as a nuchal translucency test, which were routine parts of prenatal care for women of all ages. ZQ agreed to the procedures. These tests are designed to help identify a woman at high risk for carrying a baby with Down syndrome. The test results came back with a slightly higher risk than usual for a woman of ZQ's age, and her doctor recommended she consider amniocentesis during her second trimester.

© Springer International Publishing AG 2017
L. A. Roscoe and D. P. Schenck, *Communication and Bioethics at the End of Life*,
https://doi.org/10.1007/978-3-319-70920-8_2

ZQ shared the test results with her husband, but neither of them expressed particular concern. They were young—in their late 20's—and their scientific training allowed them to think through the chances of a false positive result (where the test indicates a problem that does not in fact exist). Their graduate student health plan provided coverage for routine prenatal care but would not cover amniocentesis or any other tests available at the time (such as chorionic villus sampling or cordocentesis) that might have offered more conclusive results. ZQ's pregnancy progressed uneventfully, and she continued her dissertation research. She had arranged with her doctoral committee to take a short break after the baby's birth, and then planned to defend her dissertation in early January and travel to China shortly thereafter.

ZQ went into labor a week before her due date. Labor and delivery were routine, and she gave birth to a six-pound baby girl. The infant was diagnosed at birth with Down syndrome based on her appearance, and the diagnosis was later confirmed through chromosomal karyotyping ordered by the pediatrician. The baby also had frothy white bubbles in her mouth, difficulty breathing, and a very round, full abdomen, symptoms common in babies with tracheoesophageal fistula (TEF); an x-ray confirmed this to be the case.

TEF is an abnormal connection in one or more places between the esophagus (the tube that leads from the throat to the stomach) and the trachea (the tube that leads from the throat to the windpipe and lungs). Normally the esophagus and trachea are two separate, unconnected tubes. When a baby with TEF swallows, liquid can pass through the abnormal connection between the two, allowing the liquid to enter the lungs, which can cause pneumonia and other problems. This defect occurs in approximately 1 in 4000 births and is more common in babies who also have other birth defects, such as Down syndrome. TEF is almost always identified shortly after the baby's birth rather than prenatally, and it is considered a surgical emergency. Surgery to close the connection between the esophagus and the trachea should be done quickly after the baby is born so that the lungs are not damaged and the baby can be fed. ZQ's pediatric surgeons believed this would be a fairly simple procedure for her baby daughter and would lessen the chances of any subsequent problems such as gastrointestinal reflux disease (GERD).

ZQ was devastated by the news of her daughter's condition. She and her husband had fervently hoped for a son, so the first disappointment was their baby's sex. The diagnosis of Down syndrome was even more overwhelming. ZQ was told by the physicians that her daughter was otherwise healthy and would be able to live a fairly normal life once the fistula was repaired. The baby did not appear to have additional heart or digestive tract abnormalities that sometimes accompany TEF. The risks and benefits of the surgical procedure were explained, along with the fact that the baby would not survive without the surgery. All ZQ needed to do was provide consent for the procedure and it would be done in the next few days. She refused. Buddhist traditions led ZQ to believe that her baby's disabilities were punishment for sins in the infant's previous life. ZQ believed that her disabled daughter, whether working off her karmic burden or by disturbing the order of things by her imperfect appearance and abilities, would bring ill fortune to her family. She also knew that her daughter would never be accepted by her traditional Chinese family once she returned to China to be reunited with her husband.

A pediatric social worker met repeatedly with ZQ to convince her that her daughter's life had meaning and purpose, and that the surgery should proceed sooner rather than later. ZQ insisted she did not want the surgery to be scheduled. She wanted her daughter to be kept comfortable and allowed to die since she was "defective." An ethics consult was called, and the team recommended that all efforts be continued to convince ZQ to bond with her daughter and consent to the surgery. ZQ continued to explain her position: "This baby is not acceptable to me and will not be accepted into my extended family in China. This baby's disabilities are a punishment for her past sins, and since I am her mother, I should be allowed to decide that her life has no quality or worth. This baby should be allowed to die peacefully," she added.

The social worker brought in the parents of a child with Down syndrome, who attempted to demonstrate to ZQ that their child had a life of good quality, and that she had brought much joy to their family. The social worker and pediatrician explained to ZQ that in America the courts could intervene and take decision-making power out of her hands if she did not consent to life-saving surgery for her daughter. ZQ was unfamiliar with the idea that the court system could overrule her decision-making authority, and she was also concerned about jeopardizing her student visa status. Her husband agreed with ZQ's decisions, and reminded her of the poor ways in which disabled individuals were generally regarded in Chinese society. He also did not want the courts or the hospital to complicate ZQ's plans to complete her degree and return to China. ZQ and her husband eventually, but unwillingly, consented to the surgical TEF repair. This seemed to them to be the best course of action in order to insure the family would be reunited as soon as the baby was well enough to travel and ZQ had finished her academic work.

The surgery was successful, and during her daughter's recovery the Child Life Specialist, nurses, social worker and pediatrician spent time with ZQ to encourage her to bond with her baby daughter. ZQ seemed to acquiesce, and she left the hospital once her daughter was able to eat without complications. ZQ returned to her campus apartment, defended her dissertation research, and made plans to return to China with her daughter in order to be reunited with her husband and extended family.

Discussion Questions

1. To what extent can parents make subjective decisions about their newborn babies' quality of life and medical care?
2. The United States does not have a particularly proud history on the issue of Down Syndrome newborns, especially as related to those born with attendant esophageal problems. What do you think of the social worker's and pediatrician's "threat" to have ZQ's refusal of consent to surgery overridden by the courts? How might this have been better handled?

3. How can Western medical institutions support cultural diversity and still maintain ethical standards that reflect American cultural values?
4. When may we appropriately conclude that our ethical judgments and actions have been successful, i.e., that the safety and well-being of all concerned is insured?

A Bioethicist Responds

This case narrative is a reminder of some very dramatic chapters in the history of American bioethics. It recalls parental decisions to withhold relatively simple, yet life-saving surgeries from newborns with Down syndrome and esophageal defects (e.g., "The Johns Hopkins Baby," Gustafson 1973; and Baby Doe cases, Newman 1989), as well as the withholding of simple interventions designed to extend a newborn's expected life while markedly enhancing quality of life at the same time (e.g., The Case of Baby Jane Doe, Kerr 1984). Viewed from a slightly different perspective, the case involving ZQ might also serve as a reminder of The Case of Baby K, where a parent insisted on treatment deemed not only futile but also expensive (Annas 1994). Despite the relatively short history of bioethics overall, the American experience concerning disabled neonates could fairly said to be a complex one, wherein third parties, the courts and legislation provide a rich history (Beauchamp and Childress 2009). While none of the foregoing need be reviewed in detail, it should be sufficient to note that the case of ZQ most closely resembles the Hopkins Baby case.[1] In this case, as in many other cases, third parties have chosen to involve themselves centrally and very forcefully. One might not have expected the outcome found in this case, given U.S. history, laws, and regulations that have been established over time. Yet third parties are particularly successful at effecting the outcome in this instance if for no other reason than because the environment is not a culturally homogeneous one.

The narrative of this case offers a fairly good indication of the capacities for discernment and decision making that ZQ and her husband would be expected to possess. ZQ has made what for her is a very rational, well thought out, logical decision, given her cultural background, her religion, and the cultural tradition into which she plans to return. Viewed from her perspective, one might say that she has made a rather courageous choice. This is not to say that those of us in the west need to agree with, or embrace, her decision, but we must at least be prepared to approximate the framework of thinking from which such a decision has come.

Allopathic physicians in China offer prenatal testing to expectant mothers as they do in the west, and the available data is interesting. Zhao and colleagues report that the Chinese Birth Defects Monitoring Network (CBDMN) surveillance data for Down syndrome for the period 1996–2011 showed that the overall prenatal diagnosis of Down syndrome increased sharply from 12.98% in 2003 to 69.22% in 2011, and that the termination rate increased from 17.65 to 94.47% (Zhao et al. 2015). The authors also note that the high termination rate of affected pregnancies has led to a 55% reduction in overall Down syndrome perinatal prevalence in China. The ensuing discussion then references another recent study conducted by

Zhang and his colleagues that may be of significance regarding the problems besetting ZQ (Zhang et al. 2015). After analyzing data gathered over a period of twenty months during 2012–13, Zhang and his team concluded that it was appropriate to offer non-invasive prenatal testing as a routine screening test for fetal trisomies 21, 18 and 13 in the general population (Zhang et al. 2015). Given China's desire to limit population growth, the foregoing conclusion would appear to be strong support for such a policy. Moreover, as Deng and his colleagues mention in their article on recent trends in Down syndrome in China, the *Administrative Method on Antenatal Diagnostic Techniques* regulation issued by the Ministry of Health of China in May of 2003 encourages "all pregnant women […] to have first- and second-trimester screening and possible diagnosis for birth defects through ultrasound examination and maternal serum biomarker tests […]. Following a systemic and standardized diagnostic process, pregnancy affected by severe anomalies such as Down syndrome is allowed to be terminated at any gestational age following informed consent" (Deng et al. 2015, 312). Clearly, the Chinese of the PRC have no "official" problem with what has come to be known in the U.S. as abortion-on-demand. In fact, Deng states that most Chinese women opt to abort fetuses with malformations. Now, how likely it is that the Chinese would actually euthanize, or ask to have euthanized, critically ill newborns or those judged to be defective and therefore "unacceptable" is an open question; it is difficult to find much in the literature on this subject. Nevertheless, it is clear from the literature and available data that the advances in prenatal screening and acceptability of aborting defective fetuses in mainland China has vastly reduced the numbers of persons with genetic abnormalities, such as Down syndrome, in recent years.

ZQ suffered unnecessarily. The misfortune of giving birth to an infant with both genetic and anatomic anomalies, the first of which made the baby unacceptable to her, and, therefore, "unacceptable" not only to her but to the society into which she would soon be returning, was monumental suffering enough. The pediatric social worker and pediatrician should have been at her side providing their professional expertise in their respective roles, as they undoubtedly did, but they added greatly to ZQ's burden of suffering by bringing in the parents of a Down's child to serve as positive role models and by threatening ZQ with the possibility of a court inter- vention were she not to consent to the TEF repair. They took advantage of an already vulnerable patient, apparently made no real effort to understand her view of the entire situation, and essentially intimidated her into falling into line with their American, Judeo-Christian values. This is not to say that it would have been better had they simply acceded to ZQ's request to allow her daughter to die in the manner of the Hopkins baby, but their handling of the situation, and perhaps even the final outcome of it, was no more satisfying, no more ethically justified, than had they done so.

It would be difficult—not impossible, perhaps, but certainly very difficult—to balance the complex network of values across these two very different cultures in a way that might be acceptable both to ZQ and those providing her care. An attempt should have been made to do this, and yet there is no indication that it was. It appears that those involved in ZQ's care were concerned primarily with protecting their own western values. There is nothing wrong with that, nothing whatsoever, on the face of it. Where the health care team came up short, however, was in failing to take into account the familial and social nature of this new mother and in working with her over time through the issues caught between their two cultures. Perhaps the end result would have been no different. Yet it would have been preferable in that at least respect for the person of ZQ would have been observed rather than what could likely only be viewed as having a harsh, condemnatory decision forced upon her.

A Health Communication Scholar Responds

The most common chromosomal abnormality among Chinese babies is Down syndrome, which affects approximately one in eight hundred newborns. The incidence is nearly the same as for Caucasians in the U.S. Regardless of culture or place of birth, the chromosomal disorder carries a range of intellectual impairments and other health problems, including heart and stomach defects, a weakened immune system, poor hearing, and a shortened life span. In the U.S., it is estimated that 90% of fetuses diagnosed with Down syndrome are aborted. Abortion is generally permissible in China as well, except that sex-selective abortions are illegal. The possibilities for an individual with Down syndrome to live a life of quality and possibility, however, are quite different.

Advocates for the disabled in the United States justly celebrate the decisions of those who decide it makes sense to bring a pregnancy to term after a positive test for Down syndrome. Parents and siblings of persons with Down syndrome, as well as people with Down syndrome generally report positive experiences. In 2011, Dr. Brian Skotko, then a clinical fellow in genetics at Children's Hospital Boston, conducted a series of studies that revealed many positive aspects of lives touched by Down syndrome. Of the 2000 parents or guardians surveyed by Skotko and his colleagues, 79% reported their outlook on life was more positive because of their child with Down syndrome (Skotko et al. 2011a). Among siblings ages 12 and older, 97% expressed pride in their brother or sister with Down syndrome, and 88% believed they were better people because of their sibling with Down syndrome (Skotko et al. 2011b). A study of adults with Down syndrome found that 99% were happy with their lives, 97% liked who they were, and 96% liked how they looked (Skotko et al. 2011c). One would be hard-pressed to find these statistics in surveys of so-called healthy, normal adults.

Down syndrome is now much easier to detect than it was during ZQ's pregnancy. A blood test introduced in 2011 detects Down syndrome with 99% accuracy nine weeks into pregnancy. Parents who decide to give birth to a Down syndrome baby have time to prepare emotionally, physically, and financially for the challenges that lie ahead. For those who do not choose abortion, but feel unable to meet the unique challenges of parenting a child with Down syndrome, the National Down Syndrome Adoption Network (NDSAN) works to match children of all ages with Down syndrome with adoptive families to ensure that every child with Down syndrome has the opportunity to grow up in a loving family. The NDSAN reports that approximately 200 families are currently waiting to adopt children with Down syndrome in the U.S., many of whom have past experience with a child, sibling, or other relative with Down syndrome.

The implementation of the Americans with Disabilities Act (ADA) in 1990 improved access to services to make the lives of disabled individuals, including those with Down syndrome, and those who care for them, easier.[2] The ADA guarantees that children with disabilities cannot be excluded from public accommodations—such as private businesses, preschools, child care centers, school age child care programs, out-of-school programs, and family child care–simply because of a disability. Reasonable accommodations must be made to integrate children with disabilities into available programs, unless their presence would pose a direct threat to the health or safety of others or require a fundamental alteration of the program.

Problems remain, however. The National Council on Disability reports there are many areas in which the ADA remains unimplemented, and gaps in information, knowledge about requirements, and interest in complying with the ADA still exist. A study published in 2009 revealed a 25% decline in Down syndrome birthrates after the first President Bush signed the ADA into law (Fox and Griffin 2009). Part of the explanation lies in the unexpected messages, sometimes referred to as expressive externalities, the ADA conveyed through media and popular culture to people unfamiliar with the law. Media reports spelling out the need for new protections or the cost of implementing them can reinforce parental fears of having a child with Down syndrome.

If judged by its laws alone, China would be a global leader on disability rights. The Laws on the Protection of Persons with Disabilities, introduced the same year as the ADA, offer strong and wide-ranging protection of the civil rights of the disabled, guaranteeing employment, education, welfare and access to services. Despite the noble aspirations of the law, in practice few concessions are made for disabled individuals. Wheelchair access is non-existent, especially outside Beijing or Shanghai, and guide dogs are forbidden in most public spaces. The law says that children with special needs are entitled to proper schooling, but there are no provisions for funding such access. Local school authorities regularly turn away disabled children, telling them to go to 'special facilities' elsewhere that do not exist, or that are out of their parents' financial or geographical reach. According to a 2013 report by Human Rights Watch, 43% of disabled Chinese people are illiterate, compared with 5% of the general population.[3] Only a third received the services they need, and only a fifth got assistive devices, such as walkers, prosthetics, or adapted software; only 15% received any government funding.

The situation for disabled female children in China is worse, since they face discrimination for being female as well as being disabled. China's infamous "One-Child Policy," introduced in 1979, institutionalized the cultural preference for healthy baby boys; the 2010 census reported a ratio of 120 boys to 100 girls under the age of 5 years. In the early years of the new millennium, China's orphanages were filled with healthy girls, another reflection of the cultural preference for males and population control policies. As the rules limiting Chinese couples to one child are being phased out, parents are giving up disabled children, girls especially, because they cannot afford to care for them and services for disabled individuals are poorly structured and difficult to access (Vanderklippe 2014). As of 2014, Chinese authorities estimate that 98% of abandoned children have disabilities. Chinese traditions conflate biology and morality. The patriarchal Confucian notion of the importance of lineage sees the body, especially if male, as part of a chain of continuity stretching back to an individual's ancestors and forward to his descendants. The individual body is associated with the national body, so to have a disabled child fails not only one's family but one's country and people. A crippled or deformed body is attributed to the moral or spiritual flaws of parents, especially mothers, who failed to follow medical superstitions during pregnancy

In 1996, a series of investigative reports, including one by Human Rights Watch/Asia, and a British documentary called "The Dying Rooms," publicized the poor conditions in Chinese orphanages.[3] Staffing and funding were minimal, and foreign journalists reported that up to 90% of the baby girls who arrived at the orphanage died there (Mosher 1996; Thurston 1995). The embarrassment that ensued led to some reforms, but disabled children are still abandoned at alarming rates. Some orphanages have opened Safe Baby Hatches, places where parents can anonymously abandon children they are unwilling or unable to care for. A hatch was open for 6 weeks in spring of 2014 in the southern city of Guagzhou and received an average of five children per day. Before the hatch closed due to international outrage, 262 children, all of whom had serious health problems, were dropped off.

In late 2013 China began to relax the strict provisions of the one-child policy, partly to address the decline in the supply of labor needed to power continued economic growth. Couples can have a second child if either parent is an only child, and rural couples can have a second if their first child is a girl; further policy revisions are being contemplated. The new policy has been met with limited enthusiasm by Chinese families; only 6.7% of eligible couples in Beijing applied for permission to have a second child in the first 10 months since the rules changed (Denyer 2015). This reflects how a combination of rapid urbanization, rising incomes, and decades of government propaganda have convinced many Chinese couples that one child really is best.

These demographic patterns are further evidence that culture changes more slowly than policy reform. As health care professionals in the U.S., we can insist that ZQ make medical decisions that appear to be in her daughter's best interest. However, we can only enact a particular moral framework on our own soil, and we cannot guarantee a positive outcome for this family after their return to China. Ethical decision-making is heavily influenced by cultural notions of acceptable behavior, and there are no easy compromises between culturally opposed ideas of right and wrong. Did ZQ and her husband find the challenges of raising their daughter insurmountable and drop her off anonymously at a Safe Baby Hatch? Can we predict with any degree of certainty that the new family found sufficient emotional and financial support to raise their daughter and insure she has a good quality of life? Do we have the right or hubris to think we can change a 5000 year-old culture, one baby girl at a time? Should we try? Or in this case was the more humane, and realistic choice, to inform ZQ that a loving adoptive home would likely be found for her daughter?

Ethical decision-making also has an uneasy relationship with time. Some decisions must be made urgently, such as the decision to repair the baby's TEF. Other decisions can be made with more time for reflection and deliberation, such as whether ZQ should keep the baby or allow her to be adopted in the U.S. Decisions seem to be and are often linked—a decision to proceed with surgery with poor odds of curative success often means that difficult decisions will have to be made later about continuing or ending life prolonging treatment in the ICU. Other times, such as in this case, it might have made more sense to uncouple the decisions: encourage ZQ to consent to surgery for her baby daughter, but not link that decision to the baby's ultimate custody. Time works unevenly in helping to justify the "rightness" of the decision that gets made. At the moment of ZQ's discharge, we might congratulate ourselves on making the "right" decision in an ethically challenging situation. The baby survived, and ZQ appeared committed to parenting her. But if we extend our time frame out a bit more, to when ZQ arrives in China with her Down syndrome baby, what then? What can we make of the decisions that were made in the hospital if ZQ ends up abandoning her daughter? We can only make the best decisions possible given the information available, but in this case, ZQ's reservations and cultural restrictions were not given their due in influencing the decisions that were made.

Notes

[1]As reported in Gustafson (1973, pp. 529–30), Mr. and Mrs. X were 35 and 34 years old respectively, he a lawyer, she a hospital nurse, when she gave birth to a Down Syndrome baby with duodenal atresia. Despite the physician's attempts to reassure Mrs. X that "mongolism" was one of the milder forms of mental retardation, that "mongols" were almost always trainable and that they were famous for being happy children, she refused to grant consent for surgery to correct the intestinal block. Mr. X supported his wife's decision, and the infant was put in a side room where, over an 11-day period, the infant was allowed to die for lack of food. The hospital chose not to seek a court order to override the parents' decision.

[2]For more information, see: National Council on Disability. (2007). Implementation of the Americans with Disabilities Act: Challenges, Best Practices, and New Opportunities for Success. http://www.ncd.gov/publications/2007/July262007.

[3]For more information, see: Human Rights Watch. China: End discrimination, exclusion of children with disabilities. July 15, 2013. http://www.hrw.org/news/2013/07/15/china-end-discrimination-exclusion-children-disabilities; and Human Rights Watch/Asia. Death by Default: A Policy of Fatal Neglect in China's State Orphanages. 1996.

References

Annas, George. 1994. Asking the courts to set the standard of emergency care—the case of Baby K. *New England Journal of Medicine* 330: 1542–1545.

Beauchamp, Tom L., and James F. Childress. 2009. *Principles of biomedical ethics*, 7th ed. New York: Oxford University Press.

Deng, Changfei, Ling Yi, Yi Mu, Jun Zhu, Jun, Yanwen Qin, Xiaoxiao Fan, Yanping Wang, Qi Li, and Li Dai, Li. 2015. Recent trends in the birth prevalence of Down syndrome in China: Impact of prenatal diagnosis and subsequent terminations. *Prenatal Diagnosis* 35: 311–18. doi: https://doi.org/10.1002/pd.4516.

Denyer, Simon. 2015. Easing of China's one-child policy has not produced a baby boom. *The Washington Post*, February 6.

Fox, Dov, and Christopher L. Griffin. 2009. Disability-selective abortion and the Americans with disabilities act. *Utah Law Review*, (November 20, 2009), 845. Available at SSRN: http://ssrn.com/abstract=1362146.

Gustafson, James M. 1973. Mongolism, parental desires, and the right to life. *Perspectives in Biology and Medicine* 16 (4): 529–557.

Kerr, Kathleen. 1984. Reporting the case of Baby Jane Doe. *Hastings Center Report* 14: 13–19.

Mosher, Steven W. 1996. The dying rooms: Chinese orphanages adopt a 'zero population growth policy.' Population Research Institute. https://www.pop.org/content/dying-rooms-chinese-orphanages-adopt-zero-population-growth-policy.

Newman, Stephen A. 1989. Baby Doe, congress and the states: Challenging the federal treatment standards for impaired infants. *American Journal of Law and Medicine* 15: 1–60.

Skotko, Brian G., Susan P. Levine, and Richard Goldstein. 2011a. Having a son or daughter with Down syndrome: Perspectives of mothers and fathers. *American Journal of Medical Genetics* 155: 2335–2347.

Skotko, Brian G., Susan P. Levine, and Richard Goldstein. 2011b. Having a brother or sister with Down syndrome: Perspectives from siblings. *American Journal of Medical Genetics* 155: 2348–2359.

Skotko, Brian G., Susan P. Levine, and Richard Goldstein. 2011c. Self-perceptions from people with Down syndrome. *American Journal of Medical Genetics* 155: 2360–2369.

Thurston, Anne F. 1995. In a Chinese orphanage. *The Atlantic*, April.

Vanderklippe, Nathan. 2014. The tragic tale of China's orphanages: 98% of abandoned children have disabilities. *The Globe and Mail*, March 21.

Zhang, H., Y. Gao, F. Jiang, M. Fu, Y. Yuan, Y. Guo, Z. Zhu, M. Lin, Q. Liu, Z. Tian, H. Zhang, F. Chen, T.K. Lau, L, Zhao, X. Yi, Y. Yin, and W. Wang. 2015. Non-invasive prenatal testing for trisomies 21, 18 and 13: Clinical experience from 146,958 pregnancies. *Ultrasound in Obstetrics and Gynecology* 45(5): 530–538. doi:https://doi.org/10.1002/uog.14792.

Zhao, Weiwei, Fan Chen, Menghua Wu, Shuai Jiang, Binbin Wu, Huali Luo, Jingyi Wen, Chaohui Hu, and Shinhui Yu. 2015. Postnatal identification of trisomy 21: An overview of 7,133 postnatal trisomy 21 cases identified in a diagnostic reference laboratory in China. *PLoS ONE* 10(7): e0133151. doi:https://doi.org/10.1371/journal/pone.0133151.

Case 3—When the Family Won't Decide

TN was a baby boy born with multiple congenital anomalies, including alobar holoprosencephaly (HPE), where the brain does not develop nor divide into right and left lobes.[1] These cerebral anomalies were diagnosed in utero. He also had a cleft lip/palate, feeding problems, severe gastro-esophageal reflux (GER), failure to thrive, and seizure disorder. TN had the abnormal muscle tone and impaired motor abilities present in virtually all individuals with HPE. TN's mother (age 16) and father (age 20) lived separately with their respective parents. TN was conceived during their very brief romantic relationship, which neither desired to continue. TN's mother dropped out of high school when she found out she was pregnant, and her attempts to complete a High School Equivalency Diploma Program (GED) were not consistently productive, no doubt complicated by the birth of a seriously ill baby. TN's father had also dropped out of high school before graduating, and was working part-time as a pizza delivery driver.

During a follow-up visit to the hospital's pediatric clinic when TN was two months old, the neonatologist recommended that his parents consider allowing a PEG tube[2] to be placed to help insure TN was receiving adequate nutrition and hydration. TN's parents refused to provide consent, and his mother said he was drinking a small bottle of formula (approximately five ounces) every three hours or so; however, TN had marginal and sometimes no weight gain documented during subsequent clinic follow-up visits.

When TN was six months old, he was admitted to the hospital after a clinic visit for a three-day calorie count, which would help re-evaluate the benefits of having a PEG tube placed. The planned three-day hospital stay was extended due to TN's temperature instability, electrolyte imbalances, and increased needs for care. TN's young mother visited her baby almost every evening, but was not there in the early morning hours when her son's doctors made their rounds. His father visited irregularly and infrequently. On day eight, TN began aspirating formula, and he was transferred to the Pediatric Intensive Care Unit (PICU). The pediatric gastroenterologist strongly recommended placing a PEG tube, and after several attempts to contact TN's mother about his deteriorating condition, was finally able

© Springer International Publishing AG 2017
L. A. Roscoe and D. P. Schenck, *Communication and Bioethics at the End of Life*,
https://doi.org/10.1007/978-3-319-70920-8_3

to reach her by telephone. "I don't want my baby to have any more things done to him," she said, "We just want him back home with us." She reluctantly agreed to the PEG tube placement, and told her parents that the doctor kept saying it was urgent and that she needed to decide right away.

The PEG tube was immediately placed, but it did little to prevent TN's continued decline over the course of his hospitalization. He required intubation and became ventilator dependent, and despite the PEG tube, his feeding issues continued. He developed life-threatening anemia (hemoglobin <4; normal range is 10.5–14). The attending physician left numerous messages for multiple family members over the course of nearly 20 h, attempting to obtain consent for an emergency blood transfusion. Risk Management was contacted, and they stated that if TN's life was in danger, the indicated blood transfusion should be given unless the patient's mother could be reached and she specifically and emphatically said "No." If his mother could not be reached, or if she could not make an immediate decision, the patient should be transfused as necessary to save his life. TN's family did finally come to his bedside that evening and met with the attending physician. TN's mother agreed to a "one-time only" blood transfusion, which was ordered.

The blood transfusion stabilized the baby's hemoglobin, but the next day pediatric neurosurgery was consulted because of TN's worsening neurological status. The neurosurgeons found that TN had no gag, cough or corneal reflexes, and only minor respiratory effort. Their assessment, recorded in the baby's electronic medical record, stated: "This baby has severe anomalies with hypothalamic failure. Prognosis is dismal. The anomaly is trying to run its natural course. Intervention will be to no avail." TN was then 10 months old.

TN's mother was only minimally interactive at that point; her earlier regular visits to the hospital had become almost non-existent, and his father stopped visiting all together. The attending physician again attempted to encourage TN's parents to come to the hospital so his worsening condition could be better explained to them. After many phone calls and voice mail and text messages, TN's parents agreed to come to the hospital to meet with their son's physicians. During the initial meeting, his parents explained that TN had been doing "fine" until they were "forced" to consent to the PEG tube placement and that they believed the trauma from that procedure had caused the complications leading to TN's present decline. Subsequent meetings were scheduled, but TN's father refused to attend and would also not answer his phone. TN's maternal grandparents and maternal aunt eventually attended, along with the baby's mother, who said little. The baby's family members were antagonistic and hostile, and they continued to blame the hospital and physicians for TN's declining health and precarious status.

The physicians and nurses involved in TN's care believed that his family had never been realistic about the severity of the baby's condition. At the end of another lengthy meeting between members of the extended family and the medical team, including the palliative care service, the maternal aunt stated she felt the decisions had been taken out of their hands when TN's mother was "forced" to consent to the PEG tube; she had, after all, declined it in the past. She also reiterated that the family would have preferred for the baby to be at home, and she said they were never given that option. The aunt then said that the family would need to pray for guidance before making any decisions to limit or change TN's care to comfort measures only, despite the neurologists' note about the baby's dismal prognosis. The family members further stated that they did not believe that TN was suffering, and they said, "God would not let our baby suffer and will take him when it is time."

One month later it became clear TN was near death. Urgent calls were placed one afternoon as well as the next morning with no success in reaching any of the family members. Convinced there would be no response from them, the attending physician removed the ventilator, and TN died within minutes. The family finally came in later that day, and were given private time at TN's bedside and support from pastoral care and child life services.

Discussion Questions

1. How might better communication between the medical staff and TN's family have improved his care?
2. Did the medical team handle TN's case appropriately?
3. Under what circumstances might it be appropriate for medical staff to make end-of-life decisions for a patient whose decision makers cannot be reached?
4. How should hospitals deal with parents or other family members who refuse to take on an active surrogate decision-making role?

A Bioethicist Responds

Little TN died at less than one year of age after what must have been a very unpleasant life suffering from feeding problems, severe gastro-esophageal reflux, failure to thrive and a seizure disorder, not to mention having to undergo PEG tube placement as well as frequent heel-sticks for blood-level monitoring. Add to this physical discomfort, his parents' general unresponsiveness to calls from physicians and hospital staff, and their apparent lack of interest in working with the health care team in their son's best interests, the overall portrait of TN is a sad one indeed. One wonders if it could not have been better.

HPE is a serious condition, most often fatal, and it is important to understand the facts surrounding it. Mortality is high for all newborns with HPE, although some of them survive beyond the neonatal period, with a few even surviving many years. Alobar HPE is the severest form, but studies have shown that some 20–30% of this group may even survive beyond one year. The higher mortality rates among the alobar group tend to correlate with the severity of concomitant factors such as brain malformation, facial malformation, chromosomal abnormalities and multiple congenital anomalies. Developmental disabilities are present in virtually all persons with HPE, but they are severest in the alobar group, where children generally make little progress in development and tend to have profound global impairments. Children with HPE also commonly suffer from other problems such as hydrocephalus, seizures, motor impairment, dysphagia, pulmonary and gastrointestinal problems, and hypothalamic and endocrine dysfunction (Levey et al. 2010).

It is evident from the case narrative that TN had numerous significant deficits from birth. Without knowing more specific medical details about his case, it would be difficult to say whether he might have survived beyond one year of life, yet it is understandable why the neonatologist recommended PEG tube placement at two months, and then again at six; the fact that TN was surviving, albeit with inadequate nutrition, was probably sufficient reason for the neonatologist to want to offer him any reasonable means toward continued life on both occasions. What TN did not have in his favor, it appears, was the kind of family support that was needed. His parents refused to give consent for the PEG at two months; his mother refused again at six months when first asked to consent, but she eventually agreed "because of the urgency of the doctor's request," as she reportedly told her own parents. Once the PEG was in place, TN's mother and father visited him less and less, and they became less responsive to phone calls. When the baby's situation became critical and the parents were summoned to a meeting with the medical team, though only after repeated efforts, TN's mother and father blamed his problems on the trauma of PEG tube placement. Subsequent meetings were found to be even less successful, with the father refusing to attend at all and other family members becoming even more hostile to the medical care team. Given the attitudes and behaviors of his family as he entered his tenth month of life, TN's chances of survival certainly could not have appeared good. Furthermore, there was no promising home life into which he would be likely to return. It is little wonder that things developed as they did in the eleventh month of TN's life.

Infants as critically ill as TN simply cannot be expected to survive without proper care and support on multiple levels. High early mortality is associated with severe cases of HPE, and the case of TN would surely qualify as one of them. As Kaliaperumal et al. commented recently, "Care of children surviving with HPE requires multidisciplinary input from different medical and surgical specialties including rehabilitation to ensure optimal patient care and parental support" (2016, 808). Gupta et al., suggest that parents should be counseled prenatally about the poor prognosis of babies with HPE—only 50% of babies with alobar HPE will survive four to five months, and only 20% of those cases will live to twelve months of age. If the pregnancy is continued, the baby should be immediately referred for physical and occupational therapies (2010). Redlinger-Grosse and her colleagues have pointed out that even when HPE is detected prenatally and parents are told of the poor prognosis for their baby, it is often given as an uncertain one (2002). Likewise, they say, parents of children born and diagnosed antenatally with the condition are often told their children will die within days or weeks of birth (Redlinger-Grosse et al. 2002). Redlinger-Grosse et al., focus on the complex decision-making process that parents of a prenatally diagnosed HPE infant go through in deciding whether to continue or terminate the pregnancy, and the many informational, emotional and supportive needs parents will require in order to help ensure optimal outcomes for their HPE children. They note, in any case, that the majority of parents receiving a prenatal diagnosis of a serious abnormality terminate their pregnancies. Once again, the case narrative of TN amply demonstrates that neither he nor his parents had the kinds of support they needed, if for no other reason than his parents' refusal to accept the advice and counsel of those providing medical care for their son.

The most obvious question to be asked in this case is whether the health care team made the correct decision in removing the ventilator as TN lay near death and urgent calls to the family went unanswered. One might suggest that nothing would have been lost, and no harm done, had comfort measures been initiated, or titrated upwards, while further efforts were made to locate the mother, perhaps even with police assistance. We do not have all the medical facts of the situation at that time, however, and there may have been reasons that this approach was not feasible. On the other hand, even if it were, TN's physicians had by this time come to know enough about his parents and extended family to know that there was every likelihood of pushback, rancor, hostility, or another failed attempt at meaningful communication resulting from continued efforts, perhaps even the possibility that they would be accused of having caused the infant's crisis or of wanting to "kill" him were they to propose removing the ventilator. Perhaps they chose just to go ahead and withdraw the ventilator without securing parental consent since it seemed clear TN was nearing the end of his life anyway, and it was just as convenient the parents were difficult to reach this time?

A more likely explanation, however, and one which can be supported ethically, is that the decision to withdraw was likely made on a best interests standard, acting to do the most good for, and to cause the least amount of harm to, an infant who had already suffered greatly throughout his entire life, knowing that there was no possible way to cure or even ameliorate his situation, and knowing also that death would occur within a matter of days. There is simply no reason under the circumstances of this case to continue what has become futile care.

The only other option open to the team would have been to ask for a *guardian ad litem*, but that would not have been a wise choice, at least in this writer's opinion, for this reason: *guardians ad litem* generally act very conservatively on the part of the patients in whose interests they serve. And there is nothing at all wrong in this approach, for that is their job. This approach may not always produce the most ethical result, however, and if a guardian had been appointed in the case of TN, there is every chance he or she would have refused to consent to the withdrawal of the ventilator while yet further attempts to locate the parents were made. There would have been nothing wrong with this in principle, but if much time had passed, and the medical team had every reason to think that it could well have, the effect would only have been to add to the already great suffering of TN. It is true, in fact, that the family came to the hospital later that day, but given the way they had behaved in the past to their son, as well as toward the medical team throughout his hospitalization, the decision to withdraw life support was entirely justified at the time it was made.

A Health Communication Scholar Responds

This case involves two young people—not a couple—attempting to cope with a likely unplanned pregnancy and a very sick baby. Young people often face hardships in adjusting to parenthood, including lack of education, limited financial resources, inadequate parenting skills, unstable relationships, and transportation and housing issues. The present case adds to these common challenges the particularly difficult circumstances of coping with the severity and irreversibility of TN's medical condition.

TN was diagnosed with a rare congenital abnormality, holoprosencephaly (HPE), estimated to affect between 1 in 5000–10,000 live births. Risk factors include maternal diabetes, infections during pregnancy (syphilis, toxoplasmosis, rubella, herpes, cytomegalovirus), and various drugs taken during pregnancy (al-cohol, aspirin, lithium, thorazine, anticonvulsants, hormones, retinoic acid), but often no cause can be determined.[3] Most pregnancies with a fetus diagnosed with HPE end in miscarriage; only 3% survive to delivery and the majority of those babies do not survive past the first six months of life. It is very unlikely that TN would have lived to 11 months of age without the aggressive interventions pro-posed by the medical team, to which his young parents only reluctantly consented.

We have no information about the pre-natal care TN's mother received, or if she was exposed to any of the possible risk factors for HPE. We only have the vague comment that "cerebral anomalies were diagnosed in utero." Was TN's mother apprised of the very serious nature of these anomalies, including that they were sure to seriously foreshorten TN's life? Was she offered counseling to determine her willingness to continue the pregnancy, or given the option to terminate the preg-nancy? Surely these young parents would have benefitted from pre-delivery counseling so they would have been in a better position to contribute to decisions about fetal and neonatal management of their baby. The case narrative gives us no clues about whether or to what extent such counseling took place. This might have been the first point at which decision-making authority was transferred from these young persons to the medical team.

Given TN's likely prognosis of death within six months, the pediatric intensive care team's recommendation to place a PEG tube was curious. Even without assisted feeding, TN had outlived his probable prognosis of death within 6 months, but how much longer did the medical team expect him to live, and at what quality of life? TN's mother had refused PEG tube placement previously, and she only reluctantly agreed the second time the issue arose, when TN began aspirating formula and the pediatric gastroenterologist determined that a decision needed to be made urgently. The parents had a right at this point to refuse this intervention due to the seriousness of TN's continued medical challenges and the inability of aggres-sive medical care to address them. With the right support, they may have opted for comfort or palliative care in the PICU. Given his prognosis of likely death within 6 months, another option might even have been hospice care delivered at home, especially since the family indicated at a later meeting their desire to have TN home with them. Maybe TN's mother was hesitant to agree to the PEG because she realized that one intervention inevitably led to subsequent interventions, none of which would contribute materially to her baby's quality of life or address his underlying medical condition. Maybe the family had started to accept his limited life expectancy and his diminished quality of life. It is difficult to determine the motivations of the pediatric intensivists—did they see TN's feeding issues as just another challenge that must be met at all costs, simply because TN was a baby in the PICU as Chiswick (2008) suggested? The third time this young family faced pressure to make an "emergency" medical decision for TN involved the blood transfusion, which once again solved a short-term problem but did little more than prolong the baby's suffering. We do not know TN's mother's health insurance status, but she was young enough to still be covered under her parents' medical insurance policy. We can fervently hope that financial incentives on the part of the medical care team were not a significant factor in this case. Nonetheless, neonatal and pediatric intensive care units have become increasingly important profit centers since the 1990s (Harrison 2008; Simpson 1999; Silverman 1993). Hospital systems have become, in the words of pediatrician Lantos, "hooked on neonatology," with clear incentives to expand medical care options to this new population (2001). This fact, coupled with the advances in technology for premature babies as well as those born with disabling conditions, certainly contribute to the overall ethic of aggressive intervention that characterizes U.S. hospital care.

We want to believe that parents have the autonomy and authority to make decisions for their seriously ill children, but there is evidence that shows families often have little informed input in decisions regarding resuscitation or treatment (Harrison 2008; Keenen et al. 2005; Partridge et al. 2001). Although most of these studies concerned infants born prematurely, their findings also apply to infants born with serious disabilities. There are many players in neonatal and pediatric intensive care dramas, and the parents may be the least informed about their child's medical condition or of their legitimate medical and ethical options. A survey of 149 practicing neonatologists in New England showed more than half saw their role as providing information in a neutral manner; far fewer saw their role as helping parents balance the risks and benefits of treatment or the familial or social consequences of their decisions (Bastek et al. 2005). Counseling was found to be largely 'directive' and focused on short-term issues, and parental options to forego resuscitation or other aggressive interventions were rarely mentioned. Another study showed that neonatologists who provided information in a neutral manner left parents feeling isolated; parents preferred support and engagement with the medical care team in the decision-making process (Payot et al. 2007).

It is not surprising that there was frustration and hostility on both sides. The frustration of the medical staff is understandable. Decisions needed to be made and the rightful parental decision makers could not be reached. In the absence of consistent parental or extended family involvement, who should take on that responsibility? The family's level of involvement was just enough to make a referral to a child protective services agency unnecessary. The limited parental involvement left the medical staff to make decisions for this vulnerable patient.

From the perspective of TN's parents and extended family, it appears that several times during their baby's short life they either did not have sufficient information about their options, or decisions were forced because of an emergent medical crisis or perceived urgency on the part of the medical staff. TN's parents might have felt that since the physicians insisted on the PEG tube placement (instead of allowing a natural death at that point)—that decisions had in fact been taken out of their control. Instead of a blood transfusion, to which his mother only reluctantly agreed, why not discuss palliative care, since the very next day the neurologists documented TN's prognosis as "dismal?" During one of the last family meetings, the maternal aunt stated they wanted TN home and that they would pray over the decision to accept comfort care for TN. Why wasn't that allowed? Hospice care could surely have been provided in the home to support TN and his family, medically, psychologically, and spiritually. It seems that TN's prognosis was always poor, so it must have been confusing for these young parents to receive messages compelling them to consent to the PEG tube, and to the blood transfusion, and almost simultaneously to have his prognosis described as "dismal". And the medical staff members' decision to remove TN from life support without his family present seems particularly precipitous.

There are lessons from this sad case that might be helpful in minimizing the suffering and frustration of all involved in future similar cases. Chiswick reminds us to foreground the best interests of the infant (2008). Actions and decisions can focus on issues that infants could have an interest in, such as the degree of pain and suffering involved in their care, the futility of medical intervention, and the likelihood of survival free of serious disability. The best interests of an infant are inevitably based on the perceptions of others, including parents and medical staff, and family-centered neonatal care should be based on open and honest communication between parents and professionals on medical and ethical issues (Harrison 1993). It is not acceptable to give information vaguely or euphemistically; even young parents should not be shielded from information about uncertainties or controversies relevant to their child's care. The young age of TN's mother was a factor in this case, but it appears that she did have the support of her extended family to help interpret the information given and assess the choices to be made. Perhaps the social worker involved in this case should have visited TN's mother and her parents at their home. Hospitals, especially intensive care units, are foreign and hostile territories to those unaccustomed to their sounds, smells, and technology.

We should also expect some disconnects in terms of attributions and expectations when we confront a young family with a very difficult set of choices and decisions to make. Patients and medical professionals inhabit different social worlds. The nurses and doctors in the NICU and PICU routinely encounter babies and young children with serious, life threatening, and disfiguring medical conditions. Those outside of the medical professions are unlikely ever to be in contact with a baby as sick or as disfigured as TN.[4] The young parents here needed to be made to feel a part of the medical care team, working together with medical professionals to alleviate their baby's pain and suffering, ensuring the safety and efficacy of treatments, and learning parenting and decision making skills. Above all, parents need to have the option and support to say 'no' to burdensome treatments for their children. In this case, TN's mother should have been given a chance to decide whether to carry the pregnancy to term, as well as the risks and benefits of PEG tube insertion and other medical interventions. Neonatologists should be realistic about the resources that are or are not available to parents in decision making and caring for seriously ill newborns. The default option in the hospital setting, even more so in intensive care units, is always more treatment. In some cases, the better alternative, the one that adequately accounts for the baby's best interests and the parents' preferences, might be comfort measures, which could have been started in the delivery room if not sooner.

Notes

[1]Holoprosencephaly (HPE) is a birth defect that occurs during the first few weeks of intrauterine life. HPE is a disorder in which the fetal brain does not grow forward and divide as it is supposed to during early pregnancy (incomplete cleavage of the embryonic forebrain/failure of the prosencephalon to cleave into the cerebral and lateral hemispheres). The most severe form of HPE is alobar, where the brain is not divided and there are severe abnormalities, including absence of the interhemispheric fissure; a single primitive ventricle; fused thalami; and absent third ventricle, olfactory bulbs and tracts and optic tracts.

[2]Percutaneous endoscopic gastrostomy (or PEG) is an endoscopic medical procedure in which a tube (PEG tube) is passed into a patient's stomach through the abdominal wall, most commonly to provide a means of feeding when oral intake is not adequate.

[3]For more information, see The Carter Centers for Brain Research in Holoprosencephaly and Related Malformations (https://www.carterdatabase.org/hpe/about/).

[4]Other facial abnormalities present in many children diagnosed with HPE include a flat single-nostril nose (cebocephaly), close-set eyes (hypotelorism), or just one upper middle tooth (single maxillary central incisor). More severe facial deformities may include a single central eye (cyclopia), a nose located on the forehead (proboscis), or missing facial features.

References

Bastek, T.K., D.K. Richardson, A.F. Zupancic, and J.P. Burns. 2005. Prenatal consultation practices at the border of viability: A regional survey. *Pediatrics* 116: 407–413.

Chiswick, M. 2008. Infants of borderline viability: Ethical and clinical considerations. *Seminars in Fetal & Neonatal Medicine* 13: 8–15.

Gupta, A.O., P. Leblanc, K.C. Janumpally, and P. Tanya. 2010. A preterm infant with semilobar holoprosencephaly and hydrocephalus: A case report. *Cases Journal* 3: 35.

Harrison, H. 2008. The offer they can't refuse: Parents and perinatal treatment decisions. *Seminars in Fetal & Neonatal Medicine* 13: 329–334.

Harrison, H. 1993. The principles for family-centered neonatal care. *Pediatrics* 92: 643–650.

Kaliaperumal, C., S. Ndoro, T. Mandiwanza, F. Reidy, F. McAuliffe, J. Caird, and D. Crimmins. 2016. Holoprosencephaly: Antenatal and postnatal diagnosis and outcome. *Childs Nervous System* 32: 801–809.

Keenan, H.T., M.W. Doron, and B.A. Seyda. 2005. Comparison of mothers' and counselors' perceptions of predelivery counseling for extremely premature infants. *Pediatrics* 116: 104–111.

Lantos, John D. 2001. Hooked on neonatology: A pediatrician wonders about NICUs' hidden cost of success. *Health Affairs* 20: 233–240.

Levey, E.B., E. Stashinko, N.J. Clegg, and M.R. Delgado. 2010. Management of children with holoprosencephaly. *American Journal of Medical Genetics Part C Seminars in Medical Genetics* 154C: 183–190.

Partridge, J.C., H. Freeman, E. Weiss, and A.M. Martinez. 2001. Delivery room resuscitation decisions for extremely low birthweight infants in California. *Journal of Perinatology* 21: 27–33.

Payot, A., S. Gendron, F. Lefebvre, and H. Doucet. 2007. Deciding to resuscitate extremely premature babies: How do parents and neonatologists engage in the decision? *Social Science and Medicine* 64: 1487–1500.

Redlinger-Grosse, K., B.A. Bernhardt, K. Berg, M. Muenke, and B.B. Biesecker. 2002. The decision to continue: The experiences and needs of parents who receive a prenatal diagnosis of holoprosencephaly. *American Journal of Medical Genetics Part A* 112: 369–378.

Simpson, J. 1999. Response to 'Neonatal viability in the 1990s: Held hostage by technology' by Jonathan Muraskas et al. and 'Giving moral distress a voice: Ethical concerns among neonatal intensive care unit personnel" by Pam Heffernan and Steve Heilig. *Cambridge Quarterly Healthcare Ethics* 8: 524–526 [Book review].

Silverman, William A. 1993. Is neonatal medicine in the United States out of step? *Pediatrics* 92: 612–613 [Letter to the editor].

Case 4—Aggressive Treatment for a Child's Inoperable Tumor

RK was a four-year-old boy diagnosed with an anaplastic Wilms' tumor (also known as a nephroblastoma), a rare childhood cancer (but the most common pediatric kidney cancer). Approximately 500 new patients a year are diagnosed in the United States, and most are between the ages of 3 and 4 years old. The most common sign is a firm mass in the belly. The tumor expands slowly and is often quite large when detected, but is usually highly treatable. RK's tumor, however, was determined to be inoperable because of its massive size and because it was encasing vital organs. The anaplastic histology (AH) was also prognostically significant[1] (Beckwith and Palmer 1978; Dome et al. 2006). Only about 6–10% of Wilms tumors have the nuclear enlargement, nuclear atypia, and irregular mitotic figures that signal anaplasia, but the prognoses for patients with AH is worse than for patients with more favorable histology.

RK subsequently received aggressive chemotherapy with vincristine and actinomycin-D, and concomitant radiation to shrink the tumor. Despite these best efforts, he succumbed to respiratory distress, and experienced significant weight loss related to loss of appetite. Imaging studies confirmed tumor growth and metastases to his lungs and abdominal lymph nodes.

Upon their son's diagnosis, RK's parents did extensive on-line research about Wilms tumor. Most of the information they found indicated that the disease was highly treatable. However, when their son's physicians talked with the parents at length, over multiple meetings, they explained that there was virtually no chance of RK surviving his disease due to its advanced stage at diagnosis, the unfavorable anaplastic histology, and the metastases to their son's lymph nodes and lungs. Nevertheless, the parents insisted that the medical staff continue aggressive treatment despite knowing the high risk of failure and the harsh side effects that treatment would cause. Side effects from the chemotherapeutic drugs RK received included hair loss, mouth sores, loss of appetite, nausea, vomiting, diarrhea, infections, bruising, bleeding, and extreme fatigue. RK experienced all of these. The nurses would try to tempt RK with popsicles and other small treats, and he would routinely turn his back to them and face the wall.

© Springer International Publishing AG 2017

L. A. Roscoe and D. P. Schenck, *Communication and Bioethics at the End of Life*,

https://doi.org/10.1007/978-3-319-70920-8_4

RK's parents were of course extremely worried about their son. A phase I clinical trial for Wilms tumor patients became available, and RK's parents immediately consented to have him enrolled.[2] However, after only one experimental treatment, RK was too ill to be included in the study. Several times the parents were approached by the medical team and asked to consider changing RK's care plan from aggressive to palliative. This recommendation was not well received, and the parents instructed the medical staff "not to bring this up again." RK's parents insisted that he wanted to keep fighting. Child Life specialists and social workers were not allowed to work with RK, so it was difficult to assess what in fact RK perceived and wanted. Complicating this case was the fact that the parents were not of the same religion. RK's mother was a Jehovah's Witness, but she had agreed to a blood transfusion for RK at one point. His father tended to rely heavily on non-medical alternative approaches to health care, but in any case, the parents were united in their insistence on aggressive treatment and limited pain medication for their son.

RK's nurses reported that it was clear to them he was in a great deal of pain. When they asked him how he was feeling when his parents were out of his room getting coffee, RK said, "my tummy hurts really bad!" His parents demanded he not be given more than a patient-controlled analgesia (PCA) morphine pump adjusted to administer only the smallest doses. "We want RK to be alert and awake enough to know us and talk to us!" they explained. The parents even refused to allow their son to be given Tylenol, saying, "He doesn't have a fever; he doesn't need it." Nonetheless, it was clear to the nursing staff that RK would be more comfortable if he were given adequate pain medication.

An ethics consult was called in which the futility of continued treatment and the suffering of the child were discussed. It was apparent that the nursing staff members in particular were having a difficult time coping with RK's situation. The suffering and pain experienced by infants and young children is a significant source of moral distress for nurses and other medical staff members who are compelled to provide treatment they believe to be futile or unnecessarily burdensome (Hefferman and Heiling 1999). Lengthy discussions were held regarding the burdens and benefits of treatment, the fact that medical personnel take an oath promising to "first do no harm," and that they believed continuing to provide aggressive care for RK was indeed harming him. The moral distress described by the nurses on the unit resulted in most of them asking to be assigned to RK's care no more than one shift per week, even though many of them were quite attached to him. RK was a personable and friendly child who liked to draw pictures for the nurses and was proud to see his artwork displayed at the nurses' station.

The 90-minute ethics consult resulted in the following options for the continued care for this patient and his family: (1) continue to deliver aggressive medical care and limited pain management according to the parents' requests (and ignore the nursing staff members' increasing moral distress and RK's obvious suffering); (2) have the attending physician present a plan of care that medical staff members believed would be in the child's best interests; i.e., one that would "do no harm," and allow palliative care and better pain management, along with spiritual and emotional support; or, (3) plan for transfer of the patient to another medical care facility where the parents' demands could be better supported. The medical team met with the parents, but they remained unchanged in their insistence that full aggressive treatment and limited use of pain medication be continued. "Palliative care is for dying kids, and that isn't an option of our son! And neither is a transfer anywhere else. He's staying here and we want you to provide every treatment available to help him get better!" they said.

The parents also said their son was to remain a full code, and when he had a seizure five days later, an aggressive response was provided. He was successfully resuscitated, but died within minutes. Shortly after his funeral, the parents divorced. In retrospect, it appeared to the staff that RK was the "glue" that held the family together, and that this explained, in part, the insistence that RK continue to receive aggressive treatment.

Discussion Questions

1. Did the physicians act appropriately in acceding to the requests of the parents regarding RK's treatment? What about the question of "do no harm?"
2. Could/should the staff have attempted to obtain a court order to stop futile treatment and provide indicated pain medication?
3. Were the parents guilty of abuse of their son?
4. How much say should a minor child, even one as young as four, have in determining his care?

A Bioethicist Responds

Stories of parents disagreeing with the treatment recommendations of physicians caring for their children may be legion, but this case seems particularly troubling. One can easily imagine the parents' distress, especially after having been told of RK's grim prognosis in view of their son's advanced disease at diagnosis, unfavorable histology and distant metastases; their consent to enroll him in a Phase I trial, even though he was too ill to continue after only one treatment; and their insistence on continued aggressive care in lieu of the palliative care recommended by the medical team. Loving parents not infrequently ask that every effort be made for their severely ill child afflicted with a deadly tumor, but RK spent his last days in a great deal of unnecessary pain, serving only, as it later appeared to the medical staff, to keep his parents together. The existential pain surrounding RK's death that must surely have been felt by those who had cared for him, and who had hoped to help provide a better final chapter to his life, could only have been exacerbated upon hearing of his parents' divorce following his funeral.

Parents clearly bear the legal right and responsibility for their minor children, including the consent, or refusal of consent, to medical treatment for them. This right and responsibility may also be viewed as a freedom, as it seems to have become in the U.S., where a parent's adherence to particular religious teachings can come into play; Jehovah's Witnesses and Christian Scientists are examples of such.

Members of these faith traditions will generally opt for faith-based approaches to health care, either by consulting practitioners within their faith communities, or solely as individuals relying on faith alone, rather than for care provided by state-licensed physicians. Judges <u>may</u> grant exemptions to established law designed for the protection of minor children, such as not requiring standard and indicated medical care for a very sick child in need of that care, when parents refuse it on religious grounds, <u>if</u> the parents can show evidence to their practice of faith healing and provided this practice is established within the parents' religious tradition (Merrick 1994). The issues surrounding parental refusal of treatment recommendations for their minor children, religion, parental rights, suffering and/or child endangerment (at least from a health care team member's perspective) and law have never been adequately settled, and there is little evidence that they will be any time soon (Woolley 2005).

Hickey and Lyckholm attempted to deal with this problem through examination of several twentieth century cases, concluding that, "The right of a sick child to appropriate medical care supersedes the right of a parent to withhold that care for the sake of religious beliefs. The refusal of medical treatment on behalf of minor children is only supported when (1) the effectiveness of the treatment is in doubt, (2) the burdens of treatment outweigh its benefits, and/or (3) the care is refused by a decisionally capacitated minor" (Hickey and Lyckholm 2004, 273). However, their recommendation that members of the healthcare community inform themselves about the doctrines, previous court cases and current laws of major religious groups prohibiting medical treatment still leaves physicians, nurses and others unequipped to deal with anyone like the parents of RK whose beliefs led them to demand futile and burdensome treatment for their son. Diekema (2004) suggested that the *harm* principle might provide a more appropriate threshold than the *best interests* standard most frequently invoked to challenge parents' rights to make medical treatment decisions for their children. Other writers have attempted to offer guidance to improve communication, understanding, and trust building (Kon 2006), as well as guidelines for conflict resolution (Pinnock and Crosthwaite 2005). Yet all of these authors reluctantly concluded that should parents refusing treatment for their children prove to be totally intransigent, with the children at risk of serious harm, the only remaining option may be to address the courts. The legal system is rarely an elegant or efficient vehicle for resolving medical treatment disputes. If the court had been involved in RK's case, it is quite likely that the parents' right to demand continued treatment would have been upheld. No judge wants to limit parental decision making and wake up to see a headline describing the state's complicity in "killing" a child against the parents' wishes.

Clinicians are also certainly wary of finding themselves in adversarial positions with families, and perhaps nothing could open that possibility faster than involving the courts. Yet, as Alessandri writes, "On rare occasions, judicial involvement is necessary, and it should be undertaken as a non-adversarial process so that the possibility of rebuilding the relationship with the family remains." (Alessandri 2011, 631). There have certainly been times when a more forceful approach has been required, such as the rather dramatic case reported in Britain several years ago where a High Court judge found it necessary to make a 15-year-old girl a ward of the court in order to save her life and also to ban her mother from visiting the girl in the hospital *pro tem* until her daughter's condition had stabilized (Dyer 2014).

The foregoing offers a very brief overview of some of the most pertinent literature on this subject, but it adds little in the way of hope to the already sad narrative of RK's final days. Curiously, a very similar case occurred several years prior to RK, where a 6-year-old girl, KM, was admitted to the hospital with an advanced Wilms tumor. Her parents were advised of her terminal condition and that only palliative chemotherapy would be appropriate in her case. Her parents refused all analgesia for their daughter except for ibuprofen with a small amount of codeine with acetaminophen; they maintained that KM was not in pain and held to their concern that opioids were addictive and would cause sedation, thereby decreasing KM's ability to interact with her family. Several complex factors seemed to shape KM's parents' behaviors, including their denial of their daughter's prognosis and their misperception of opioid therapy (Weidner and Plantz 2014). KM's parents held fast, refusing to allow the administration of opioids for pain until her last hour of life, when she asked for medicine to help her breath and said she was ready "to go to heaven." Her parents then agreed to a small dose of an anxiolytic that provided sedation, and she died soon thereafter. Weidner and Plantz conclude their piece, as have others, that in the absence of a negotiated compromise, the invocation of Diekema's *harm principle* and the enlistment of a child protective agency (i.e., the Courts) is the way to go. Still, one cannot help but wonder why either KM's or RK's terrible agonies had to go on for as long as they did.

We know a bit more here about the case of RK; perhaps, therefore, it is easier to understand how the complex dynamic that developed between the parents, their son and his oncologists as one group, and how other groups such as palliative care, Social Work and Child Life, came to be seen as adversaries of the first group, most especially by RK's mother. She is easily perceived as a fearful, aggressive, manipulative woman capable of taking control of an entire healing enterprise that is oriented toward maximizing the good of the sick person, and her primary focus is upon her own needs such that she succeeds, ironically, in ensuring that far more harm than good is brought to her son. Laying blame solely at her feet for the sad ending to RK's story would be fruitless, however, particularly since she was never going to be swayed; she simply would not engage in order to avoid discussion entirely, or she would respond in passive-aggressive ways, as she did after the formal ethics consult, which was just another way for her to maintain control over the entire situation. The mother in this case was most likely unaware of what was really going on inside her, and in any case the real locus of control was not within her but rather within the health care professionals surrounding her and her family.

The problem with those health care professionals, however, is that they did not take proper and effective control of RK's situation at an appropriate moment. This should probably have been done at the time the mother refused to allow any further conversation about palliative care and/or most certainly when she began insisting on the limited dosage of pain medication for RK. It is understandable that one medical practice is unlikely to intrude on the "turf" of another if they are ever to function well again on other cases, so palliative care would not have been expected to challenge pediatric oncology. There are perhaps others in the hospital who could have intervened when it became apparent that things were simply not going well for a particular patient, and among them are none better equipped than nurses.

Nurses generally feel they are powerless, especially in the presence or in the shadow of physicians, whose orders and instructions they have been taught to follow explicitly, and who, they have also been taught, outrank them. Likewise, their professional mores include holding family members in the same respect as patients. Nurses today are frequently the ones to call ethics consults, but they are not the ones likely to demand much more than that if they see situations like RK's that warrant further attention. Any member of the health care team should be able to request a palliative care consult; such consults should not be able to be refused by either family members or the attending physician. Nurses should feel greater empowerment; they should be made to feel that they will be respected for the medical/scientific expertise that they possess, and they should not be afraid to speak up as advocates on behalf of patients. Nurses should be assured that if they act and speak truly for the patient's good, they will be supported for it, even if they are the ones to push for serious discussion or reluctantly ask that the Courts become involved. This should obtain for other members of the health care team as well, be they therapists, technicians, aides or whoever may have integral roles to play in the welfare of patients. Creating a safe environment wherein this kind of conversation can happen will undoubtedly call for something of a culture change in many health care institutions requiring significant time and effort, but such time and effort might be remembered as minor investments if the number of cases with outcomes like KM and RK declined appreciably.

A Health Communication Scholar Responds

The death of this little boy, about whom we know little other than that he was very sick and liked to draw pictures, was complicated by obvious and apparently insurmountable communication difficulties. There are other troubling aspects of this case as well.

It seems possible that there was a delay, maybe a considerable one, between the time RK's parents might have felt something was wrong with their son and when they sought medical counsel. In his recent book about his daughter Zoe's diagnosis and treatment for Wilms tumor, Elisha Cooper recounts that the day after he discovered a lump on Zoe's side, he and his wife took her to New York-Presbyterian Hospital for immediate diagnosis and treatment (Cooper 2016). RK's mother was a Jehovah's Witness and his father was described as preferring natural or alternative methods of medical care. When they discovered the lump on RK's side, did they first try herbal or alternative treatments, or did they ignore it and hope for the best? Did it go unnoticed for some period of time? Perhaps they immediately sought medical treatment despite having different preferences for their own health care and it was unfortunately the case that RK's cancer had already metastasized. The case narrative makes it clear that both parents were united in their insistence that RK continue to receive aggressive treatment, including remaining a full code, despite whatever other differences of opinion might have characterized their relationship, parenting styles, or willingness to seek medical treatment.

It is also noted that RK's parents did extensive on-line research about Wilms tumors after their son was diagnosed. Wilms tumor was named after Max Wilms, a German doctor who wrote one of the first medical articles about the disease in 1899.[3] In Zoe Cooper's case, an imaging study revealed the Wilms tumor on her kidney to be Stage III, and even though there were complications during her surgery, which required the removal of the tumor, her kidney, and part of her colon, the surgeon reassured Cooper and his wife that Zoe had "a good cancer." Zoe's parents were reassured that after 22 weeks of chemotherapy and targeted radiation therapy it was "going to be okay." Searches for information about Wilms tumor almost immediately present the good news about long-term survival rates. According to the St. Jude Children's Research Hospital website[3] long-term survival rates are excellent: about 85–90% of Wilms tumor patients with *favorable histology* (like Zoe) can be cured. Cure rates for patients with *anaplastic histology*, (like RK), a more aggressive form of Wilms tumor, are lower. Dome and colleagues reported survival estimates for patients with Stage I anaplastic Wilms tumors as 69.5% (four-year event free survival) and 82.6% (overall survival). For patients with favorable histology, survival estimates rose to 92.4% (four-year event free survival) and 98.3% (overall survival) (Dome et al. 2006). Were RK's doctors able to describe the significant difference tumor histology made in survival rates? Despite the medical team's best efforts, perhaps RK's parents chose to believe instead the hopeful news about Wilms tumors on various pediatric cancer websites and therefore insisted on treatment on their son's behalf. And what are we to conclude about RK's parents' insistence on limiting his pain medication, despite his obvious discomfort?

Bluebond-Langer (1978) wrote an insightful book based on her dissertation research with leukemic children that provides a useful theoretical framework for beginning to understand RK's parents' insistence on continued aggressive treatment. Bluebond-Langer would likely say that RK's parents knew their son was dying, and equally important, RK knew he was dying. RK knew his parents knew, and his parents thought RK probably knew too (but they hoped he didn't, since no one had told him). And both parties were determined not to let on to the other. Bluebond-Langer described this as *mutual pretense awareness*, a concept borrowed from Glaser and Strauss (1965), who described how the behavior of a dying patient, including his or her interactions with others, could best be understood by the "awareness context" in which it took place. An awareness context is the context within which people interact, including what each knows of the other's situational cognizance and status.[4]

Bluebond-Langer's work with leukemic children demonstrated that mutual pretense awareness was the dominant mode of interaction between the dying child and his or her parents because it offered a way to do what society expected of them. The child could act as if he or she had a future, for which children are supposed to prepare. Parents could maintain their responsibilities for protecting and nurturing, which become very difficult in the face of a child's terminal illness. Their decisions not to reveal their awareness to one another reflected their knowledge of the social order into which they had all been socialized. The child and parents used mutual pretense to protect one another—the dying child did not want to make his or her parents sad or give up on a future, and the parents did not want to feel they had not adequately protected their child.

Bluebond-Langer demonstrated how dying children gradually became aware of their changed relationship to their social worlds, including their relationships with their parents. The children she observed and talked with passed through stages of understanding their disease and how sick they were, and eventually understood they would not get better and would die, even if in the past they had relapsed and recovered. The children she observed would make references to their impending deaths through their conversations and play in the presence of medical personnel or non-family members, but would maintain an impenetrable silence with their parents. Once the child recognized death was imminent, most did not play with toys very often, and when they did, their play usually involved references to death and disease. For example, they would put dolls or toy animals in graves. Sedentary activity, like coloring, increased, but the range of themes expressed tended to decrease. Most pictures dealt with destruction, storms, fires, graveyards, and religious images, even among children who were not raised in a religious faith.

Children's comments about the future also changed once they had this awareness of their impending death. For example, one 5-year-old boy wanted his Christmas presents in October, and in June several children asked whether they could prepare to go to school in September. Even the doctors in Bluebond-Langer's study who doubted children could know their prognosis without being told thought these actions showed the children were suspicious, perhaps even probing for information, about how long they had to live. Some dying children also started to refuse to comply with procedures and tests, and when the nurses would relent, that further reinforced the children's ideas of how close they were to dying. How could medical procedures be overlooked, when formerly they were so important to the child's recovery?

RK and his parents appeared to conform to the contextual boundaries and rules of mutual pretense awareness, one of which is to avoid "dangerous" topics, including the child's level of pain. More pain means sicker, and sicker means closer to dying. RK may have felt comfortable telling the nurses that he was in pain when his parents were not in his room, but admitting that in front of his parents would spoil the illusion that he was feeling no different than he was previously, when he was not as sick. Similarly, insistence on continued treatment also signaled no change in prognosis; continuing treatment as usual signaled RK's status had not changed.

It is also quite possible that RK's parents acted selfishly in insisting on prolonging his life, and that their consistent refusal of pain medication for their son was to enable him to interact with them so they would not have to interact with one another. Surely the finality of losing a child, especially if his death might also end a marital relationship, is something most people would want to postpone. Guilt over delaying treatment or imposing their own preferences for medical treatment on RK (to the extent that either happened) might also have motivated decisions to do everything possible to prolong their son's life. We cannot know for sure, but the concept of mutual pretense awareness offers at least another possible explanation for otherwise inexplicable behavior. We do not know the inner dynamics of families whose every interaction and decision occurs under the harsh glare of the medical gaze. We can, however, continue to advocate for the well-being of patients, while simultaneously extending forgiveness and understanding to parents and family members who do not always act in ways that minimize our moral distress or conform to our ideas of a good death.

Notes

[1]Anaplastic is a term used to describe cancer cells that divide rapidly and have little or no resemblance to normal cells. In 1978, Beckwith and Palmer published a detailed histopathologic review of Wilms tumors that were collected on the first National Wilms Tumor Study (NWTS-1). Approximately 6% of the tumors had cells with nuclear enlargement, nuclear atypia, and irregular mitotic figures, and were considered to have anaplastic histology (AH). The presence of anaplasia was prognostically significant; 44% of patients with AH died as a result of disease, whereas only 7.1% of patients without anaplasia, the so-called favorable histology (FH) subtype, died as a result of disease.

The more recent fifth National Wilms Tumor Study of 2596 patients with Wilms tumor who were enrolled onto NWTS-5, found 11% of tumors to have anaplasia (AH). Four-year event-free survival (EFS) and overall survival (OS) estimates for assessable patients with stage I AH (n = 29) were 69.5% (95% CI, 46.9–84.0) and 82.6% (95% CI, 63.1–92.4). In comparison, 4-year EFS and OS estimates for patients with stage I favorable histology (FH) were 92.4% (95% CI, 89.5–94.5) and 98.3% (95% CI, 96.4–99.2) (Dome et al. 2006).

[2]Clinical trials are conducted in a series of steps, called phases, each of which is designed to answer a separate research question (Mahipal and Nguyen 2014).

Phase I Researchers test a new drug or treatment in a small group of people for the first time to evaluate its safety, determine a safe dosage range, and identify side effects.

Phase II The drug or treatment is given to a larger group of people to see if it is effective and to further evaluate its safety.

Phase III The drug or treatment is given to large groups of people to confirm its effectiveness, monitor side effects, compare it to commonly used treatments, and collect information that will allow the drug or treatment to be used safely.

Phase IV Studies are done after the drug or treatment has been marketed to gather information on the drug's effect in various populations and any side effects associated with long-term use.

[3]For more information, see St. Jude Children's Research Hospital website: https://www.stjude.org/disease/wilms-tumor.html

[4]Bluebond-Langer's work is based on Glaser and Strauss' earlier work that described how the behavior of the dying patient, particularly in interaction with others, can best be understood in terms of the "awareness context" in which it takes place. An awareness context is "what each interacting person knows of the patient's defined status, along with his recognition of the others' awareness of his own definition.... It is the context within which these people interact while taking cognizance of it" (Glaser and Strauss 1965, 10). Glaser and Strauss identified four types of awareness contexts, one of which is mutual pretense awareness.

References

Alessandri, Angela J. 2011. Parents know best: Or do they? Treatment refusals in paediatric oncology. *Journal of Paediatric and Child Health* 47: 628–631. https://doi.org/10.1111/j/1440-1754.2011.02170.x.

Beckwith, J.B., and N.F. Palmer. 1978. Histopathology and prognosis of Wilms tumor. *Cancer* 41: 1937–1948.

Bluebond-Langner, Myra. 1978. *The private worlds of dying children*. Princeton, NJ: Princeton University Press.

Cooper, Elisha. 2016. *Falling: A daughter, a father, and a journey back*. New York: Pantheon.

Diekema, Douglas S. 2004. Parental refusals of medical treatment: The harm principle as threshold for state intervention. *Theoretical Medicine and Bioethics* 25: 243–264.

Dome, Jeffrey S., Cecilia A. Cotton, Elizabeth J. Perlman, Norman E. Breslow, John A. Kalapurakal, Michael L. Ritchey, Paul E. Grundy, J. Marcio Malogolowkin, Bruce Beckwith, Robert C. Shamberger, Gerald M. Haase, Max J. Coppes, Peter Coccia, Morris Kletzel, Robert M. Weetman, Milton Donaldson, Roger M. Macklis, and Daniel M. Green. 2006. Treatment of anaplastic histology Wilms' tumor: Results from the fifth national Wilms' tumor study. *Journal of Clinical Oncology* 24: 2352–2358. https://doi.org/10.1200/JCO.2005.04.7852.

Dyer, Clare. 2014. Judge makes 15-year-old girl ward of court after mother refuses to allow nasojejunal tube feeding. *British Medical Journal* 348: g2566. https://doi.org/10.1136/bmj.g2566.

Glaser, Barney G., and A. Anselme Strauss. 1965. *Awareness of dying: A study of social interaction*. Chicago: Aldine.

Hefferman, Pam, and Steve Heiling. 1999. Giving 'moral distress' a voice: Ethical concerns among neonatal intensive care personnel. *Cambridge Quarterly Health Ethics* 8: 173–178.

Hickey, Kenneth S., and Laurie Lyckholm. 2004. Child welfare versus parental autonomy: Medical ethics, the law, and faith-based healing. *Theoretical Medicine and Bioethics* 25: 265–276.

Kon, A.A. 2006. When parents refuse treatment for their child. *JONAS Healthcare Law, Ethics, and Regulation* 8: 5–9.

Mahipal, Amit, and Danny Nguyen. 2014. Risks and benefits of Phase I trial participation. *Cancer Control* 21: 193–199.

Merrick, Janna C. 1994. Christian Science healing of minor children: Spiritual exemption statutes, first amendment rights, and fair notice. *Issues in Law and Medicine* 10: 321–342.

Pinnock, Ralph, and Jan Crosthwaite. 2005. When parents refuse consent to treatment for children and young persons. *Journal of Paediatric and Child Health* 41: 369–373.

Weidner, Norbert J., and Diane M. Plantz. 2014. Ethical considerations in the management of analgesia in terminally ill pediatric patients. *Journal of Pain and Symptom Management* 48: 998–1003. https://doi.org/10.1016/j.jpainsymman.2013.12.233.

Woolley, S. 2005. Children of Jehovah's witnesses and adolescent Jehovah's witnesses: What are their rights? *Archives of Disease in Childhood* 90: 715–719.

Case 5—Is There Life After Death? A Case of Post-mortem Sperm Retrieval

5

AB and MB grew up in the same suburban neighborhood. They played together as children, along with their siblings and other neighborhood friends. It was a friendly block, and several families socialized regularly, including AB's and MB's parents. AB and MB were high school sweethearts who broke up when they both went away to college and moved away to start their respective careers. When they were in their early 30's, the two reconnected at a family wedding and were soon planning their own wedding; both families were overjoyed to hear this good news. This would be the first marriage for both, and neither had prior living children.

Shortly after they married, MB stopped using contraception in the hope of becoming pregnant. After attempting to conceive a child unsuccessfully for approximately two years, AB, a healthy 35-year-old, and MB, a healthy 33-year-old, made an appointment to consult with a fertility specialist. The physician recommended a number of standard fertility tests, and test results indicated multifactorial infertility without a clear cause. AB had some abnormal semen analysis parameters, although he had an adequate number of healthy sperm. He had no surgically correctable fertility issues. MB had also been evaluated and found to have no underlying fertility disorder that could be surgically corrected. She was started on cycles of fertility drugs to try to stimulate healthy ovulation. AB had not been advised to bank any sperm nor had MB had any eggs harvested since they were only at the early stages of considering additional fertility treatment such as in vitro fertilization (IVF).

AB frequently rode his motorcycle, whether it was to and from work, to run errands, or merely for pleasure in his free time; he was an experienced rider. One evening, however, he was involved in a crash and suffered severe head trauma even though he was wearing a helmet. He was noted to have few other injuries upon admission to the emergency department, but he required intubation and was admitted to the ICU. It was eventually determined that he would have no meaningful recovery from his brain injury, and he was subsequently found to be brain dead by the physicians caring for him. This was carefully explained to MB. As AB had no advance directive nor had he ever appointed a health care surrogate or

© Springer International Publishing AG 2017
L. A. Roscoe and D. P. Schenck, *Communication and Bioethics at the End of Life*,
https://doi.org/10.1007/978-3-319-70920-8_5

durable power of attorney, his wife automatically became his proxy decision maker.[1] MB decided that her husband would not wish to continue life-prolonging treatment under such circumstances and that it should therefore be withdrawn.

She did, however, have one request prior to withdrawal. MB stated to the physicians that she would still like to have a biologic child by her husband; they had wanted "desperately" to have a family and had tried hard, albeit without success. She asked if it were possible to harvest sperm that could then be used later for IVF. A social worker interviewed both AB's and MB's families independently, each of which confirmed that the couple had been undergoing fertility treatments and that AB had indeed wanted to be a father. The hospital's ethics committee met to discuss the case and ultimately recommended that MB's request to have her husband's sperm harvested posthumously be honored.

The urology service was then consulted. The chief of urology discussed the options for sperm retrieval with MB, who then gave formal consent for an orchiectomy (removal of testes) to be performed. The reproductive specialist was present, who confirmed receipt of live, healthy sperm as determined by microscopy. AB's sperm were immediately cryopreserved, with the cost of this process paid for by MB. Life support was then withdrawn from AB, and he expired within minutes. MB was naturally grief-stricken by this series of events, and she was provided all appropriate care by family services and pastoral care. At some point, however, either before or after AB expired, she revealed to someone on the medical staff that she knew her husband was having an affair with a woman in a nearby town. "I loved him so much, and I think our fertility issues were a factor in my husband's affair. I knew about it but I never confronted him, but I think he was on his way back from seeing his lover when he got in the accident," she tearfully confided.

Almost two years passed with MB having made no attempts at in vitro fertilization with AB's sperm. She began dating another man and was seriously considering marriage. When MB and her new partner discussed the possibility of having children, she said she very much wanted to be a mother. She told her boyfriend that she had experienced some fertility problems in the past, but she did not mention her late husband's banked sperm. MB was troubled about what to do with AB's sperm, and having the issue unresolved was an impediment to her moving into a more lasting committed relationship with her new partner.

MB had a younger sister who had been living with a same sex partner in a deeply committed relationship for almost 9 years. The two women had discussed having children, and had agreed that MB's sister would be the birth mother if a suitable sperm donor could be found. MB's sister had been a part of the neighborhood friends along with AB, and AB remained close friends with MB's sister and her partner when he and MB were married. MB encouraged her sister not to wait too long to attempt becoming pregnant given the fertility issues she had experienced, and then offered her late husband's sperm to her sister and her partner. MB's sister was enormously grateful, and was pleased that the sperm donor was someone with whom she had been very close. Both MB and her sister hoped that this offer would allow both couples to move forward with their lives and still maintain a connection with AB.

Discussion Questions

1. What should be considered when performing elective procedures on patients who cannot consent?
2. Should MB's sister accept the donation of AB's sperm from MB?
3. If MB's sister decides not to use AB's sperm for the purpose of assisted reproduction, for what other purpose(s) could AB's sperm be used? For example: as a general sperm donor, for sale or gift to someone else? For use in research?

A Bioethicist Responds

This case is interesting enough had it ended simply with MB requesting AID (Artificial Insemination by Donor) to begin the family she and AB had intended to raise together, albeit as a single parent. If MB had chosen instead, for whatever reason, to use AB's cryopreserved sperm to attempt a pregnancy after her remarriage, the ethical issues in that case would certainly have been more complicated. Yet, the situation as presented in the case narrative above leaves us with a most extraordinarily challenging ethical issue to address. This couple had attempted conception without success for about two years, they had consulted a fertility specialist, and they had been found to have "multifactorial infertility without a clear cause."[2] MB had also begun taking fertility drugs to stimulate healthy ovulation. Unfortunately, however, AB's tragic motorcycle accident, his physicians' eventual determination that he would have no meaningful recovery from his traumatic brain injury and their ultimate declaration of brain death jointly raised the first serious ethical issue in the case: the question of the permissibility of harvesting AB's sperm prior to withdrawal of life-support.[3]

Postmortem sperm retrieval (PSR) was first reported in 1980 (Rothman) and there has been considerable discussion of the ethics surrounding the issue since that time. Debate has included topics such as the interests of the child conceived; consent; the inferred wishes of the donor patient; questions as to whether physicians must honor requests for PSR; questions related to the storage of sperm; paternity; and, inheritance. An excellent review of these issues (Strong et al. 2000) obviates even a general overview of them here; it is useful to note only that while Strong and his colleagues do not view a premortem consent, either verbal or written, as an absolute requirement for PSR, they clearly do support physicians acting positively on a PSR when it is made by a legal surrogate, provided there is substantial evidence that the patient himself would have approved of the request (emphasis added; Strong et al. 2000, 745).

Any request for a PSR by a surrogate has to deal with the issue of consent before the request can be honored. MB's initial query as to the permissibility of harvesting sperm from her husband prior to withdrawal of life support from him thus immediately raises the issue of respect for AB as a person incapable of giving consent to the procedure. We have already established that MB was his legally authorized representative and is, therefore, authorized to consent to medical procedures on his behalf. MB and her husband had clearly wanted to have a family, having invested time and money into the effort; independent testimony was collected from the families of AB and MB to the effect that both of them had wanted children; and the social worker had documented comments from family members to the effect that AB had wanted to be a father. That data speaks collectively and directly to AB's intentions and values, leaving no question as to the legitimacy of MB's request for sperm retrieval. However, the ethics of the request is still left somewhat open to question for we have no information of any kind to suggest that AB had ever even discussed the idea of PSR with MB. This is not unusual; even if AB had had an advance directive the issue of PSR is not a standard question or prompt on any advance directive templates with which we are familiar.

We turn then to the issue of presumed consent. An excellent summary of opinions to date on presumed consent for PSR, including those from the medical, legal and academic communities, can be found in the work of Tremellen and Savulescu (2015). These writers take up arguments both for and against presumed consent for posthumous sperm retrieval and assisted conception, but they, themselves, concluded it to be entirely justifiable. They went even further to suggest that "presumed consent" should become the default for allowing posthumous conception (Tremellen and Savulescu 2015, 10). Central to their argument is their view that the welfare of the child is the most important consideration of all, yet the welfare of the child is precisely the issue that evokes the next ethical question regarding MB's request: MB's true motivating reason for wanting to become pregnant and raise a child by herself. If she wants a child because of the plans that she and AB had had all along, that may be one thing. But, if her desire for a child has primarily to do with keeping the memory of AB alive in another person, she may be acting disproportionately out of selfish motives. The work of Orr and Siegler (2002), as well as that of Hostiuc and Curca (2010), offers very good supporting discussions in this regard.

This still does not clarify concerns about MB's motivations, however, should we feel that necessary in assessing the ethics of her request, but we will pass on this here for two reasons: (1) there simply is no way to assess further MB's true motivations, given the data available to us, and (2) more compelling ethical questions remain to be examined in this case, beginning with MB's offer of AB's sperm to her sister. MB's sister expressed enormous gratitude at this offer as well as her desire to accept it, and one must either accept or reject that comment as valid, before moving on to the ethics of the gift itself. We will choose to accept the sister's comment at face value, again for two reasons: (1) there is absolutely no indication that MB's sister might have feelings in any way contrary to what is expressed in her statement; and, (2) MB knew that her sister and her partner, who had been living together in a deeply-committed relationship for almost nine years, had agreed that LC would become the birth-mother for their children whenever they might be able to identify a suitable sperm donor.

Can MB just "gift" AB's sperm to her sister, wonderfully selfless and generous as that may seem? A search of the literature offers little help in this regard, with the exception of an article that reported the results of a survey and content analysis of medical institutions with posthumous sperm procurement (PSP) programs and protocols (Bahm et al. 2013). A supplemental chart to this article indicates there are several institutions with established protocols stipulating that sperm procured posthumously may be used only by the wife or partner of the donor, and that the sperm is nontransferable and may not be donated or sold (Bahm et al. 2013, 10–11). The American Society for Reproductive Medicine's Ethics Committee opinion on posthumous collection and use of reproductive tissue also clearly supports the position that only the surviving spouse or life partner of the decedent should have any ethical claim to preserved tissue, and that programs offering these services should decline requests for them from other individuals in the absence of written instructions from the decedent (Ethics Committee of the American Society for Reproductive Medicine 2013, 1844).

Still, none of the foregoing need mitigate against MB making the gift to her sister as she does. In fact, her gesture at the time might be quite understandable. As she seriously considers remarriage, she no doubt views AB's preserved tissue through a different lens than she once did. There is also an argument to be made for her act as an expression of genuine love, the offer of a precious gift to a close family member struggling with a rather unique reproductive issue. If her sister does accept the gift all appropriate testing would need to be done prior to implantation to protect the health of all parties concerned. And, one should have every reason to expect a child resulting from this union to receive every bit as much love, care and nurturing, and to be as happy, psychologically well-integrated and ultimately well-adjusted as any child born into a "traditional" family through sexual intercourse. No significant data to the contrary have been produced. Serious questions would certainly be put forth regarding this child's identity, however, such as: Whose son/daughter is he/she, really? Will he/she be told who the sperm donor was, and why or why not? What are the implications of this in either case? Should the son/daughter be told about AB, and if so what would the relationship between them be like? How important, in fact, is the issue of identity (i.e., knowing exactly who one's biological parents are)? A good deal has been written on these issues over the years, and it can be said once again that there is no consensus on these difficult questions. One very solid point of agreement, however, and one that could be seen to anchor virtually all else, is that which focuses on the well-being of the child (The President's Council on Bioethics 2004).

This is one of those cases about which it may truly be said that there is no right or wrong answer; some answers may just be better than others. Readers must decide for themselves as to which of the possible answers they might envision may be better than others, keeping in mind the needs and interests of all parties concerned, while also attempting to maximize good, minimize harm and protect the dignity of all.[4]

A Health Communication Scholar Responds

The earth belongs in usufruct to the living: the dead have neither powers nor rights over it.
—Thomas Jefferson, in a letter to James Madison, dated Sept. 6, 1789.[5]

Post-mortem sperm retrieval (PMSR) used to be sensationalized by the press and accompanied by court room drama. Prior to his death by suicide, William Everett Kane, a prominent California attorney, deposited sperm at a cryobank in Los Angeles. He left instructions that his sperm should be used by his girlfriend of five months, Deborah Ellen Hecht (Hecht v. Superior Court 1993). Kane's college-aged children went to court to request that the sperm be destroyed; Ms. Hecht wanted the sperm released to her or her physician. After lengthy hearings, the court ultimately distributed 20% of the sperm to Hecht, and 40% to each of Kane's existing children (the same percentage by which Kane's other assets were distributed). Hecht did try to conceive, but was never able to, partly due to the decreased quantity of sperm (Williams 2011). In another situation that occurred in July, 1994, 22-year-old Emanuele Maresca was killed in a car accident 16 days after his wedding. His 22-year-old wife, Pamela, asked that post-mortem surgery be performed to extract and preserve sperm, to be used in the future to conceive a child (Stiteler 1994). Fifteen months later, Pamela had changed her mind, but was contemplating allowing her mother-in-law or sister-in-law to attempt to conceive a child with her late husband's sperm and a donor egg. Their story was covered in newspapers and the tabloid press, and the young widow and her mother-in-law were invited to tell their story on several television programs that featured sensational stories such as theirs.

As with many medical procedures, what once was rare and exceptional eventually becomes unremarkable, but the ethical and communication issues surrounding the practice are far from settled. The present case presents a multi-part ethical dilemma, each step of which contains possible communication difficulties. There is the initial and urgent decision about whether AB's sperm should be retrieved and banked, which is dependent upon whether his wife can supply convincing evidence of AB's desire to become a father. Second, when and under what circumstances can the sperm be used in an attempt for MB to conceive a child? Evidence of a desire for paternity does not necessarily mean AB would have wanted his wife to conceive, bear, and raise a child without him—either as a single parent or perhaps with another partner. Third, what rights does a child conceived in this way have to know the facts of his or her conception, and what rights in a material sense does this child have to his or her biologic father's property or estate? And lastly, does MB as caretaker of the sperm have authority to use it for any other purpose, i.e., for her sister's use?

The first and last parts of this equation are the most straightforward. AB and MB were pursuing fertility testing and possible treatment, which is clear evidence of AB's desire for fatherhood. The conversations between MB and the physicians taking care of her husband seemed to go well—decisions were made to withdraw life support and retrieve sperm based on the evidence MB had about her husband's wishes for both end-of-life care and potential fatherhood. Thus, AB's sperm can be retrieved and cryopreserved without any violation of ethical principles. Some physicians see this as an act of compassion and a way to help the grieving widow in a time of great stress and sadness (Laborde et al. 2011); others as an extension of the dead man's autonomy by enabling the birth of his genetic offspring (Williams

2011). Intent can be implied since there is evidence that AB wanted to have children with his wife, but unfortunately died before MB became pregnant (Cannold 2004). Since there is compelling evidence that AB did wish to become a father, it is also quite clear that if the evidence provided by MB to justify sperm retrieval, i.e., that AB wished to father a child with her, then AB's sperm is only to be used to father a child with MB, and MB does not have the authority to re-gift her late husband's sperm to her sister.

Let's now turn to the thornier parts of this case scenario. Can the patient's attempts at fertility treatment while alive create the reasonable inference that he would have wanted his wife to conceive, bear and raise his child after his death, possibly with another partner? The literature on this topic indicates that since sperm retrieval is time-sensitive, if there is convincing evidence of a desire for paternity, it should be allowed. But more restrictions might well be placed on the actual use of the sperm—guidelines for the use of sperm acquired under these circumstances might require psychological counseling for the wife, a one year waiting period, an assessment of the family's stability and financial status, a plan for how to tell the child about how he or she was conceived, and medical discussions about the assisted reproductive techniques to be used (Tash et al. 2003). Here there is some evidence of a breakdown in communication—the ER physicians and urologists were comfortable proceeding with sperm retrieval, but were leaving these other difficult conversations to others to address with MB. Alerting MB to the possible barriers to the future use of her late husband's sperm would have been an additional act of beneficence on the part of AB's medical team. Sperm banks, like all legitimate medical facilities, have policies and procedures that govern their care for patients and preserved tissues. It is possible that some sperm banks would be comfortable using the sperm in any way dictated by MB, but it is also likely that some sperm banks would only allow AB's sperm to be used by MB in an attempt to conceive. Being the responsible party for AB's sperm has moral and emotional consequences for MB, and the physicians involved in AB's end-of-life care and sperm retrieval should have at least mentioned them to his surviving spouse. It is curious that MB did not also discuss organ donation with her husband's medical team. An otherwise healthy young man with a traumatic brain injury, although a tragedy to his family, is the ideal candidate for organ donation. An opportunity was missed to discuss organ donation with AB's wife and family (Desai et al. 2004; Kahan et al. 1999). It is possible that such discussions took place, but they were not mentioned in the case narrative.

Should MB proceed with assisted reproduction using her dead husband's sperm, and successfully conceive and bear a child, the issues of a relationship with the child's paternal relatives, along with any inheritance or other property rights, would have to be sorted out. There is little in the probate literature on how to best sort this out, and it is usually decided on a case-by-case basis. The states that have addressed post-mortem conception have utilized the Uniform Probate Code, the Uniform Parentage Act, and the Uniform Anatomical Gift Act (Williams 2011). Surely the child should know the truth of his or her origin, and the grandparents would have a legitimate claim to visitation with their deceased son's offspring. However, family composition changes over time, and is accompanied by changes in relationships. The good relationships that presently exist between MB and her in-laws can fracture in the future, especially if she pursues a committed relationship and parenthood with another partner. Continuation of close supportive relationships between in-laws can happen, but is rare and unpredictable (Finch and Mason 1990). One has only to reflect on the high level of tension and strife between Terri Schiavo's husband, Michael, and her parents, the Schindlers. When Terri and Michael first married, Terri's parents were extremely close to the young couple, and lent them money and even allowed them to live with them temporarily. The closeness persisted between Michael and his in-laws after Terri's injury, but became contentiousness when Terri's prognosis worsened, and Michael was awarded a medical malpractice award on behalf of his debilitated wife; Michael and the Schindlers fought bitterly over Terri's care and the disbursement of those funds (Roscoe et al. 2006). The acrimonious relationships between the two parties became a defining characteristic of this case.

The few studies to be found in the family communication literature examined how divorce affected individuals' relationships with their former in-laws. Duran-Aydintug conducted in-depth interviews with former spouses and their former in-laws, and concluded that the quantity and quality of the interaction between former spouses and their former in-laws depended on the quality and quantity of this relationship before the separation or divorce (1993); we might conclude that the same dynamic would pertain in the situation of the death of a spouse. If the relationship was strong and positive pre-divorce, it might continue after divorce. If the spouse was the only link between them, relationships are likely to deteriorate.

Divorced persons who consider former in-laws as relatives will likely maintain higher levels of contact and utilize those persons as a source of support (Serovich et al. 2008). At present, the lack of family conflict surrounding the issue of sperm retrieval bodes well; it appears that many attending physicians would see conflict between a surviving spouse and the dead man's parents over this issue as a reason not to proceed with sperm retrieval (Whitney and Mian 1998).

The existence of AB's suspected girlfriend has no direct bearing on the ethics or communication issues in this case, nor does the same-sex relationship of MB's sister or the sister's desire to become pregnant with donor sperm. Since MB has already offered the use of her late husband's sperm to her sister, they are likely to have a difficult conversation in the future, since it is very unlikely that the sperm bank would be willing to release the sperm for use by AB's sister-in-law.

Notes

[1]Althougth the terms "proxy" and "surrogate" decision maker are sometimes used interchangeably, a surrogate decision maker is one appointed by the patient; a proxy decision maker is one appointed according to applicable laws when the patient has not appointed a surrogate.

[2]Both AB and MB underwent medical examinations and testing procedures in efforts to determine the likely source(s) of their difficulties in conception. An introduction to this subject may be found in *Diagnostic evaluation of the infertile female: a committee opinion* (Practice Committee of American Society for Reproductive Medicine) and *Diagnostic evaluation of the infertile male: a committee opinion* (Practice Committee of American Society for Reproductive Medicine), both published in 2015.

[3]Note, however, that there is no ethical issue here regarding MB serving as her husband's proxy decision maker. This case occurred in Florida (where the spouse of an incapacitated patient is automatically first in line as the patient's legal medical surrogate), AB and MB were legally married, and there was no indication or suggestion of a possible conflict of interest. Furthermore, MB's decision "that her husband would not wish to continue life-prolonging treatment under such circumstances and that it should therefore be withdrawn" would almost surely have been grounded on the basis of substituted judgment. A decision made on this basis is entirely justifiable and ethical, provided it is made by a person with knowledge and understanding of the values and beliefs of the patient, and who makes the decision he/she believes the patient would make, based upon his/her knowledge of the patient's values and beliefs.

Although there is no written confirmation of AB's end-of-life treatment preferences, and no mention of posthumous sperm retrieval, there did not appear to be any reason to questions MB's judgment or decisions regarding her husband's medical treatment.

[4]Other ethical issues could well be discussed here such as: the wisdom of raising a child in a single parent household; the happiness and well-being of a child raised by one parent; children raised by same-sex couples. Issues likely to evoke religious and/or political views in defense of their ethical positions, however, have been purposely avoided here.

[5]Thomas Jefferson, 7 Jefferson's Works 454 (Monticello ed. 1904), in a letter to James Madison, dated Sept. 6, 1789. Quoted in Kerr (1999). Post-mortem sperm procurement: Is it legal? *DePaul Journal of Health Care Law, 3,* 41–78.

Usufruct is a civil law term referring to the right of one individual to use and enjoy the property of another, provided its substance is neither impaired nor altered.

References

Bahm, Sarah M., Katrina Karkazis, and David Magnus. 2013. A content analysis of posthumous sperm procurement protocols with considerations for developing an institutional policy. *Fertility and Sterility* 100: 839–843. https://doi.org/10.1016/j.fertnstert.2013.05.002.

Cannold, L. 2004. Who owns a dead man's sperm? A sad outcome, but the right one. *Journal of Medical Ethics* 30: 386. https://doi.org/10.1136/jme.2003.004853.

Desai, Ravi V., Mahesh Krishnamurthy, Harish Patel, and David N. Hoffman. 2004. Postmortem sperm retrieval: An ethical dilemma. *The American Journal of Medicine* 116: 858 [Letter to the editor].

Duran-Aydintug, Candan. 1993. Relationships with former in-laws: Normative guidelines and actual behavior. *Journal of Divorce & Remarriage* 19: 69–82. https://doi.org/10.1300/J087v19n03_05.

Ethics Committee of the American Society for Reproductive Medicine. 2013. Posthumous collection and use of reproductive tissue: A committee opinion. *Fertility and Sterility* 99: 1842–1845. https://doi.org/10.1016/j.fertnstert.2013.02.022.

Finch, Janet, and Jennifer Mason. 1990. Divorce, remarriage and family obligations. *The Sociological Review* 28: 219–246. https://doi.org/10.1111/j.1467-954x.1990.tb00910.x.

Hecht v. Superior Court, 20 Cal. Rptr. 2d275, 276 (Cal Ct. App. 1993).

Hostiuc, Sorin, and George C. Curca. 2010. Informed consent in posthumous sperm procurement. *Archives of Gynecology and Obstetrics* 282: 433–438. https://doi.org/10.1007/s00404-010-1475-4.

Kahan, Steven E., Allen D. Seftel, and Martin I. Resnick. 1999. Postmortem sperm procurement: A legal perspective. *The Journal of Urology* 161: 1840–1843.

Laborde, E., J. Sandlow, and R.E. Brannigan. 2011. Postmortem sperm retrieval. *Journal of Andrology* 32: 467–469.

Orr, R.D., and M. Siegler. 2002. Is posthumous semen retrieval ethically permissible? *Journal of Medical Ethics* 28: 299–302.

Roscoe, Lori A., Hana Osman, and William E. Haley. 2006. Implications of the Schiavo case for understanding family caregiving issues at the end-of-life. *Death Studies* 30: 149–161.

Rothman, C.M. 1980. A method for obtaining viable sperm in the postmortem state. *Fertility and Sterility* 34: 512.

Serovich, Julianne M., Steven F. Chapman, and Sharon J. Price. 2008. Former in-laws as a source of support. *Journal of Divorce and Remarriage* 17: 17–26.

Stiteler, Rowland. 1994. The donor is a dead man: Long after Manny Maresca's death from a brain injury, his sperm could give life. His young widow is willing to carry their baby—and so is his 41-year-old mother. *Sun Sentinel*, November 6, 1994.

Strong, Carson, Jeffrey R. Gingrich, and William H. Kutteh. 2000. Ethics of postmortem sperm retrieval: Ethics of sperm retrieval after death or permanent vegetative state. *Human Reproduction* 15: 739–745.

Tash, J.A., L.D. Applegarth, S.M. Kerr, J.J. Fins, Z. Rosenwaks, and P.N. Schlegel. 2003. Postmortem sperm retrieval: The effect of instituting guidelines. *The Journal of Urology* 170: 1922–1925.

The President's Council on Bioethics. 2004. *Reproduction and Responsibility: The Regulation of New Biotechnologies*. Washington, D.C.: Government Printing Office.

Tremellen, Kelton, and Julian Savulescu. 2015. A discussion supporting presumed consent for posthumous sperm procurement and conception. *Reproductive Biomedicine Online* 30: 6–13. https://doi.org/10.1016/j.rbmo.2014.10.001.

Whitney, Susan, and Patricia Mian. 1998. Life after death? Ethical questions raised after a request for postmortem sperm retrieval in the emergency department. *Journal of Emergency Nursing* 24: 492–494.

Williams, Devon D. 2011. Over my dead body: The legal nightmare and medical phenomenon of posthumous conception through postmortem sperm retrieval. *Campbell Law Review* 34: 181–204.

Part II
Decision-Making: Families in the Mix

Our ethical principles and processes are predicated on respect for persons, and a call to honor the autonomy of competent adult patients. Patients who have the cognitive ability to understand, communicate, and decide for themselves can make whatever decisions about medical care that they want to: Physicians and other medical care providers are obligated to honor even treatment preferences they feel are unwise, including those that will lead to the patient's death. Honoring autonomy even means in some cases that physicians and others end up providing care that they feel is futile, that is, care that has no medical benefit, that might have an effect on the body, but cannot offer a benefit to the person.

These situations are complicated in and of themselves, but additional complexity arises from the fact that most patients are not alone in their decision-making with their physicians, but are accompanied by family members (loving and otherwise) who have their own preferences, opinions, values, and decision-making power. The threat of legal action looms over situations in which a physician refuses to comply with a family members' wishes, even if they contradict what the patient may have told the physician. Patients are not usually isolated individuals who solely take into account their own preferences when deciding about medical care. They are concerned naturally about the effect of their decisions on the loved ones who will soon be left behind to mourn their deaths and agonize about whether their loved one had "a good death."

The cases in this section concern issues such as parents who refuse to accept brain death; the ways the media influence medical decision-making; the distress caused by missed diagnoses; a family's refusal to accept palliative care for their dying son; decisions to remove life support from young patients; how codependence between a patient and family can create obstacles to healing; and situations in which a patient's stated treatment preferences or best interests are not aligned with the opinions of family members.

A short summary of each of the six cases in this section is given as follows:

Case 6—What is the Standard of Care for a Corpse?

A mother inadvertently gave her young daughter expired insulin and told the Emergency Department physicians only that her daughter was experiencing flu-like symptoms. She refused to allow her daughter to be removed from life support even after she was declared brain dead, and threatened the hospital with a lawsuit for failing to accurately diagnose her daughter's condition in a timely way.

Case 7—When the Palliative Care Team Got Fired

A 17-year-old African American man was admitted to the hospital with Stage IV metastatic colon cancer, a recurrence from 5 years previously. His mother was very religious and was at his bedside constantly, and answered all questions directed to her son by the medical team. She had faith in God and in her son's oncologist, who she believed previously cured her son's cancer. The palliative care team was forbidden to enter the patient's room after they described his extremely poor prognosis and the possibility of transitioning to comfort care only.

Case 8—A Young Woman's Wish to Die

A 19-year-old Hindu woman was admitted to the PICU of a large urban hospital after an automobile accident. She was ventilator-dependent and quadriplegic and had clinically significant depressive symptoms. The patient's uncle was the family spokesperson and declared that all life support must be stopped since the young woman's injuries were a result of karma from a past life.

Case 9—When Parents Contest an Adult Child's Advance Directive

The patient was a 25-year-old woman admitted to the hospital from hospice care with end-stage cardiomyopathy. She had an advance directive which specified a limited trial of ventilator support, which the patient's schizophrenic mother successfully contested in court.

Case 10—Please Stop Torturing Me—Unless my Wife is in the Room!

A 55-year-old well-known attorney was diagnosed with B cell lymphoma. The aggressive treatment caused the cancer to go into remission but led to multiple system organ failure. The patient's second wife insisted on continuing treatment, which included dialysis and other invasive procedures, partly because the oncologists told her that her husband's cancer had been cured. The intensive care team (and sometimes the patient) saw continuing treatment as futile, but were not able to convince the patient's wife to agree to any other goals of care.

Case 11—Who Should Make Treatment Decisions for a Battered Spouse?

The healthcare team struggled with how to reconcile American law, Western medical ethics, and their own cultural values when a woman who was a Chinese immigrant was brought to the hospital after being brutally beaten by her husband, and her husband was permitted, at least initially, to act as her proxy decision-maker.

Case 6—What Is the Standard of Care for a Corpse?

HT was a 13-year-old girl who had been diagnosed with Type 1 diabetes when she was 5 years old. Her blood sugar levels had been well-maintained by frequent daily monitoring of her blood sugar and treatment with a daily injection of insulin glargine (Lantus), a long-acting insulin, and injections of lispro (Humalog), a rapid-acting insulin, before meals. HT's doctor was currently trying some modifications to this regimen, since hormones can affect insulin requirements, and HT had recently begun menstruating. One day after school, HT complained of fatigue and went to lie down on the couch. An hour or so later, HT's mother went to check on her and observed her lethargy and slurred speech. Her mother was concerned about HT's worsening symptoms, and her daughter's doctor advised that HT should be taken to the Emergency Department immediately. On the way out the door, her mother grabbed a syringe and the Humalog, thinking that HT might need another injection if they had a long wait ahead of them.

HT's mother told the triage nurse that her daughter appeared to have "flu-like" symptoms, but was an otherwise healthy child whose diabetes was well-controlled. The nurse and ER physician decided further tests were indicated despite the mother's insistence that HT merely had the flu and her doctor was being overly cautious. HT had a fever of 102° and was lethargic upon presentation. HT was in the ED for about 2 h while tests were done, and during that time she lapsed into unconsciousness from which she ultimately never recovered. Tests revealed that HT was in a diabetic coma. HT's mother reacted to this unwelcome news with disbelief. "I haven't missed a dose of insulin in 8 years!" she exclaimed, and produced the Humalog as evidence of her preparedness. The ER doctor examined the medication and said, "You do know that this prescription has expired, right?" HT's mother reached for the bottle and said, "It's YOUR fault she's in such bad shape! If you had made an accurate diagnosis two hours ago, my daughter would be just fine!" HT was admitted to the intensive care unit where she was intubated. The next day she had a gastrostomy tube inserted to provide artificial food and fluids.

It soon became apparent by electroencephalogram (EEG)[1] that HT had no brain activity whatsoever. The hospital's critical care team informed the mother of her

© Springer International Publishing AG 2017
L. A. Roscoe and D. P. Schenck, *Communication and Bioethics at the End of Life*,
https://doi.org/10.1007/978-3-319-70920-8_6

daughter's condition and explained that in their medical opinion she would never recover. They recommended to her mother that she think seriously about with-drawing all treatment. "I refuse to stop any treatment, it's your emergency room doctors who are at fault here, and I refuse to cover for their mistakes! I insist that everything be done for HT, and if you refuse, you will have the courts and the newspapers to answer to!" she said angrily. HT's mother was a woman of strong will, given to intimidation to get what she wanted. She was separated from her husband, HT's father, who was far more passive than his ex-wife. All efforts on the part of the hospital to bring the mother to an accurate understanding of the situation were for naught.

The primary care of HT was being provided by her long-standing pediatrician. He also refused to consider withdrawal of treatment, and it became apparent with time that he would follow the mother's lead in decision-making. The hospital asked for the opinion of a team of pediatric intensive care physicians at a large children's hospital in a neighboring city. Their opinion also was that HT had no brain activity and that she would never recover, but they stopped short of saying that HT was dead by current legal, medical and ethical guidelines. HT's own pediatrician and her mother were not persuaded by this information, and the hospital continued to provide care for HT while trying to decide what to do. The case was referred to the hospital's ethics committee, which acknowledged HT's brain-death and recommended withdrawal of life support. Again, the mother refused. The hospital administrator was uncomfort-able with the idea of withdrawing treatment despite the long-established ethical principle that where medical treatment was judged to be futile it could be withdrawn (nothing is quite as futile as providing life support for a dead body). He consulted the state's attorney for an opinion and was told that should he decide to withdraw treatment the hospital would not be held liable for the girl's death according to the laws of the state, based upon all objectively obtained data and medical opinion regarding the cause of HT's condition and the impossibility of recovery. Still, the administrator was reluctant to make the decision to withdraw treatment, especially since the case had now captured the attention of the local media, which accused the hospital of wanting to "kill" HT. Once more the team of pediatric intensive care physicians was called in to give an updated opinion. Again, they reported as before. By now six weeks had passed since HT had come to the Emergency Department.

The entire hospital was now feeling the strain of the case, and almost all of the medical and nursing staff believed HT's death should be acknowledged and life support discontinued. The strain was particularly severe for the intensive care nurses who cared for HT around the clock. Some even felt offense at what they were asked to do, stating they had been trained to care for sick persons, and what in effect they were doing now was caring for a corpse. They believed this was a travesty and the body of HT was being abused. Some of them were so upset that a staff psychiatrist was asked to provide counseling for them. Another consult was sought from the ethics committee, which again recommended withdrawal of all treatment. The mother was now threatening a lawsuit.

The hospital felt it had every legal and moral right to discontinue medical support from HT's corpse and send the parents, strong-willed or not, on their way.

However, in this case, another course of action was put into play: The hospital explained to HT's mother that they could no longer keep HT on the ward due to her brain-dead status. Attempts were made to encourage HT's mother to understand that a declaration of brain death amounted to legal death, but they were not successful, likely due to the series of misunderstandings that preceded that announcement. What the hospital would do instead, at their own expense, was to set up a fully functioning ICU room in HT's mother's house, complete with round-the-clock nursing care and monitoring. HT's mother agreed, and the transfer to the home setting occurred. The hospital staff could now return to more normal operations, without the protestors, news coverage and stress that the case had engendered.

Discussion Questions

1. Does HT's mother understand and accept her daughter's diagnosis and prognosis? What factors in the case might complicate her understanding of her daughter's condition?
2. Should the hospital withdraw medically futile treatment over the objections of the mother, assuming she might sue the hospital in any case?
3. Should the media's characterization of the hospital figure in whatever decision is made?

A Bioethicist Responds

This was an extremely difficult case, though that may not be fully evident from reading the text as presented. When the case occurred, approximately 20 years ago, critically important medical and ethical issues seemed clear, but they became complicated because of actions on the part of individuals central to the case that could not easily be ignored or dismissed. The complicating factors had largely to do with HT's mother and her pediatrician of many years, their attitudes, behaviors and actions, and the resultant effects that these in turn had on other hospital staff, the community, and ultimately the hospital administrator. These effects were significant enough, in fact, that legitimate medical and ethical decisions were abrogated.[2]

HT's mother, as noted in the case, was defensive, blamed the hospital for her daughter's situation, refused to agree to withdrawal of treatment, was strong-willed, was given to intimidation in order to get her way, was head of the household (she was divorced and had custody of her children), apparently had taken the lead in medical decision making away from the family pediatrician, and eventually threatened a lawsuit. She appeared to be a skillful manipulator, and in this observer's view she managed over a relatively short period of time to take complete control of the situation and everyone involved. Her doing so traumatized virtually everyone, while also turning an admittedly very sad and tragic case that might have seen a simpler conclusion in a calmer, more rational environment into one where medical, ethical, social, psychological, emotional, legal, and public relations issues became unnecessarily, yet seemingly inevitably and inextricably, entangled.

In taking the lead in medical decision making away from her pediatrician, the mother effectively made him compromise, if not outright abandon, his responsibility as a physician. No one is ever obligated to follow the advice of their physician, of course, but in this case the mother quickly set this capable professional aside, making it clear she would sanction no treatment recommendation that did not match her idea of what was appropriate for her daughter. Thus, HT's established pediatrician did not second anything recommended by outside experts, even though the medical facts in the case were clear. She, therefore, effectively nullified his expertise as a physician despite having entrusted him with the care of her daughter in the past.

This mother, quite naturally suffering enormous grief over her daughter's situation, and perhaps some measure of guilt as well, might also have felt helpless. It seemed apparent to those who had contact with her throughout her daughter's hospitalization that she was in charge of all domestic issues related to the family. Undoubtedly she now felt that she had also to be in charge of everything related to her daughter's hospitalization. She must have seen her daughter and herself as caught in a desperate situation, that she simply had to remain in control of things, that nothing could be allowed to slip out of her control, and the idea of her daughter as already deceased was just not a rational possibility. Her demanding, bullying, and threatening behaviors, which grew increasingly difficult to deal with, appeared to support the foregoing suggestion regarding her need for control as well as the view that any different assessment of her daughter's situation on her part would render her unable to cope emotionally or to manage daily responsibilities. At the point, then, that she is in denial, that she cannot deal with reality, that she can no longer accept the judgments of medical experts and has, therefore, effectively placed her own judgment over theirs, other serious problems arise.

This is where the ethical, social/psychological/emotional, legal, and public relations issues become entangled, and it all effectively surfaced when the stress of the ICU nursing staff, who felt they were being required to provide care for a corpse, came to the attention of the ethics committee and the hospital administrator. This not only raised ethical issues related to such things as appropriateness of care, respect for persons (including the deceased), justice and allocation of resources, to name a few, but it also raised very serious, pragmatic ones for the hospital administration in terms of human resources (e.g., human resource counseling) and budgets. The mother's difficult behavior caught the attention of local radio, television, and newspaper reporters who set up shop in front of the hospital, complicating things even further.

This particular hospital was then, and remains today, a strong, vital, and critically important element of this moderately-sized city; the hospital and the city have each depended heavily upon one another for many, many years, and each has done its part to ensure the growth, development, and flourishing of the other. Thus, when pickets paraded in front of the hospital with signs pleading, "Don't kill HT," or reporters asked of hospital spokespersons if it were true that doctors had recommended withdrawing life support from HT, whether they were going to do it or not, would withdrawing life support "kill" her, or would she "die" if they did so, it only added further to the problem.

At a distance, and certainly on paper, this might not seem to be a difficult problem to work through and resolve. After all, HT, most unfortunately, is dead by all accounts. We read in the case report that the critical care team of the hospital had told the mother that her daughter was brain-dead early on, and we know that pediatric specialists brought in from another hospital confirmed that to be the case during two separate consults weeks apart. Physicians and hospitals are not required to provide futile care, and the state attorney had told the administrator that the hospital would not be held liable for HT's death if a decision was made to withdraw all treatment.

So, given that the administrator had established medical and legal principles on his side, given that it would be the ethically appropriate thing to do to withdraw treatment from HT's body, given that doing so would free the ICU staff from performing what they genuinely believed to be a travesty with regard to their professional duties and ethics, given that this would then allow for a fairer and more judicious use of limited health care dollars at the hospital, and given the hospital's already good relations with the community, along with the ability of most persons of good will to understand a properly presented explanation for the doctors' and hospital's decision to withdraw life support, why not, therefore, proceed as humanely and deliberately as possible? It would seem that on so many levels this would be the right thing to do. It was entirely justified to stop treatment from a medical point of view; virtually every principle of ethics, as well as numerous virtues, could called upon to support this action not only out of respect for the patient but with regard to the interests of justice and the well-being of hospital staff as well; HT's mother was just as likely to sue the hospital whether treatment was withdrawn from her daughter or not, but the hospital already knew that it had the law on its side.

However, all of this remained complicated by the public relations piece, a factor that seems on the face of it to challenge the patience of the ethicist, he who could logically and clearly construct an argument for withdrawal based upon points already presented. Yet all the "right reason" and proper application of medical and ethical principles, including judiciously chosen virtues, along with an appeal to the good will of the community, were still inadequate to resolve this case because of a potential public relations nightmare. Again, this had been created by HT's mother who had managed to take total control of every aspect of this situation, albeit unjustifiably, and whether anyone liked it or not. While we might imagine, or might like to imagine, public relations issues to be among the lesser factors in biomedical ethical cases generally, the mother in this case ironically turned the public relations issues into a major factor.

The mother should never have been in control, and neither should public relations issues be driving the hospital administrator's decision making. Yet, this administrator realized that it was not merely an issue of standing on principle, "doing the right thing" while also resolving his internal problem of the stress on the ICU staff, effecting a more judicious use of resources, and eventually working through some initial bad publicity towards repair of the hospital's image and respect, for it would serve him much better in the long run if he could get HT and her mother out of the hospital quickly *without* having to challenge the latter, an action that would only have served to inflame things even further, and needlessly. He thus addressed all of these issues simultaneously and ultimately resolved this very difficult case by dealing with it not head on, but rather by approaching it obliquely, finding a creative way to work with the mother's irrational view of things regarding her daughter without in any way compromising his own principles or beliefs, by pretending for example that HT was *not* brain dead, setting up an ICU with 24/7 nursing care at hospital expense in the home, relieving problems back at the hospital, and completely defusing the media problems that had been gaining traction.

This case demonstrates how a direct approach may not always be the way to proceed in seeking a path to resolution, but rather how one must sometimes find different approaches, including attempting to approximate the mind-set, world view, or mental framework with which a difficult party may be operating without compromising one's own principles or values, or without being duplicitous.

A Health Communication Scholar Responds

HT's case brings into focus parental rights and accountability in medical contexts, the power of personality to highjack ethics, and the relationship between unwanted media attention and a hospital's reputation and integrity in the community it serves. It also highlights the difficulties the language of brain death sometimes presents for family members trying to understand and make decisions about their loved ones' medical care.

The information provided in the case description reveals that HT's mother caused her daughter's medical distress by providing her with outdated insulin. We cannot know if this was an accident, or if the reason was negligence, ignorance, poverty, or disinterest. Regardless of motive (or lack thereof), HT's mother was in the uncomfortable and terrifying position of needing to secure emergency medical care for her daughter. She needed simultaneously to minimize her own culpability while maximizing her ability to make medical decisions for her seriously ill daughter. The mother's administration of outdated insulin coupled with her mis-leading account of HT's symptoms were likely enough evidence to remove HT from her mother's care and perhaps even to bring charges of child abuse or neg-ligence, or even ultimately, homicide. However, in the press of a child's medical emergency, it is often the case that we assume that parents will act in the child's best interest, and in true emergencies there is often not time to investigate the availability and suitability of alternative decision makers.

HT's mother is described as having a "strong will" and used to "getting her own way." These personality characteristics might have been exaggerated in attempts to direct attention away from her own actions and to the potential for legal action or at least a "difficult" family situation. In such cases, the natural response is to slow everything down—no health care facility or physicians' practice wants to spend time on allegations of negligence and the bad publicity and endless court cases that might result, especially since in this case there would be no continuing concern about HT's well-being since she had unfortunately died. Physicians might be tempted to back away in the face of a strong parental figure, but in cases such as this, strength might need to be countered with firmness. No need given the present circumstances to remind the mother of her culpability, but perhaps there was an opportunity to assert that the medical team was attempting to partner with HT's parents to bring about the best possible resolution to this tragic set of circumstances.

Hospitals and physicians are under no obligation to provide medical care that they determine to be futile. In HT's case, the issue of futility is even less relevant, because it appears from the case description that HT was actually dead. If we are to interpret "no brain activity whatsoever" as meaning that HT's entire brain, including the brain stem, had ceased to function, then HT was brain dead, which, according to the Uniform Definition of Death in the U.S., means that she is legally dead.[3] Did HT's mother understand that her daughter was dead? HT might have been declared brain dead after appropriate testing and examination, but she looked no different than she did the day before when she was apparently "brain alive." The respirator still inflated her chest at regular intervals, the feeding tube supplied nutrients and water, and HT continued to "rest" peacefully awaiting her much wished-for recovery.

Doctors sometimes make the mistake of asking the "permission" of family members before disconnecting life support from patients who are determined to be brain dead. This is not necessary and provides additional distress to families who may feel burdened by this decision. What might be helpful here is a "brain death" protocol in use by some organ and tissue recovery agencies. Family members are first reassured that everything that could have been done was done in an attempt to prevent their loved one's death. The "absence of brain activity" should be further clarified as brain death—our legal, moral, and medical definition of death—from which no recovery is possible and for which no treatment need be continued. The patient is dead unequivocally by virtue of meeting these specific clinical standards and need not be "declared" dead by her physicians. This is not a decision that the physicians have reached; rather it is a statement that the patient meets the legal and ethical criteria for death. The family members should be invited to witness the examination and apnea testing that determines brain death, and in the best cases, a discussion should follow about exactly when "life" support will be discontinued.

It is essential to establish that the patient is dead, that his or her status is not dependent upon a physician's "declaration" of brain death, but by the unfortunate fact that the patient has indeed died, despite everyone's best efforts to avoid this outcome. Physicians and other health care providers cannot afford to stand behind euphemisms: The patient is dead by all legal, medical and moral means, despite the fact that their appearance remains unchanged because of the life-sustaining support they continue to receive. And then there has to be follow through, despite any objections or threats of legal action, to do what is required. It must be clearly communicated to HT's parents—both of them—that HT is dead and will no longer be receiving medical care. A dead body must be treated with dignity, but is not entitled to medical support. The moral distress of the staff, especially nurses, must also be acknowledged and remediated. Nurses are not trained or required to provide medical care to corpses, and the point at which HT transitioned from patient to corpse should have been acknowledged by all who participated in her care.

However, in the case of HT, there was ambiguity about whether or when the appropriate examinations and testing of brain function had been done, and how this information had been discussed with HT's parents, not only her strong-willed mother, but also her father, who despite his diminished status as a marital partner or custodial parent, remained HT's father with rights and feelings of his own. Whatever legal action pertained must be endured, despite the cost in dollars and reputation that might ensue. The hospital had an obligation to its staff and to the dignity of this deceased young girl to do what best honored the dignity and humanity of their patient and medical staff, even if the parents disagreed. Perhaps the parents did not know their daughter's status, and that, of course, should have been clarified immediately.

The hospital, no matter the timing of the discontinuance of life support from HT, is not responsible for her death, no matter what lurid headlines might result from subsequent legal maneuvering. So what is this hospital to do? HT's mother was unable or unwilling to acknowledge her part in her daughter's death, and the hospital chose to turn a humanitarian blind-eye since HT was dead and her future parental supervision was not an issue. The very real threat of bad publicity and legal action hung over the hospital, as did the real human costs of the moral distress experienced by the medical and nursing staff.

Case Resolution. We wish we could report that once mother and daughter were in the privacy of their own home, without the specter of parental guilt, blame, and dishonesty that were unfortunately factors in HT's fate, that her mother was able to recognize not just the futility but the reality of HT's condition and request that all life support be discontinued. However, this was not the case. HT's mother filed a malpractice lawsuit and the hospital offered her $1,000,000 if she promised no further claims would be made, which she accepted. HT continued to receive nursing care in her home at the hospital's considerable expense.

Initially, HT did not appear different when brain-dead than she had previously. The ventilator, feeding tube, blood pressure augmentation, and hormone regulation to support gastric, kidney and immune function continued to do some of the things HT's body had done when alive. The ketoacidosis that HT experienced because of her poorly managed blood sugar affected many of her organ systems. The resulting acidification of her blood disrupted the membrane potential of excitable cells, particularly those of the heart and nervous system. Eventually, the heart damage and acidification of the blood were enough to cause her heart to stop beating, regardless of the ventilator and other support. At some point, HT's mother ceased to be convinced by these futile efforts to sustain her daughter's body, and agreed to the discontinuation of care.

It is not clear whether the hospital did the "right" thing here—should HT's mother have been prosecuted? Should scarce resources have been spent to create an

ICU in HT's childhood bedroom? What is clear is that there is a powerful need for clear, unequivocal communication about what death looks like in a modern ICU where machines can simulate nearly every function of a living child, and where fears of liability and bad publicity make us all afraid of naming what we see in front of us and taking appropriate action. Doctors need better ways to discuss what death looks like in the ICU with family members, including when it is and is not appropriate to ask for permission to remove "life" support.

Notes

[1]Electroencephalogram or EEG is a test that measures and records the electrical activity of the brain. It is used for many purposes, such as confirming a diagnosis of epilepsy; and it is also used to determine if a person who is in a coma is brain-dead.

[2]David P. Schenck was a member of the hospital ethics committee where this case occurred. The case attracted the attention of national news outlets, though only briefly. Some of what is discussed here was not widely reported at the time, although care has been taken to reveal nothing that would violate standard HIPAA regulations or that was not generally known locally at the time.

[3]The Uniform Determination of Death Act (1981) states: "an individual who has sustained either (1) irreversible cessation of circulatory and respiratory functions, or (2) irreversible cessation of all functions of the entire brain, including the brain stem, is legally dead."

Case 7—When the Palliative Care Team Got Fired

The patient, BK, was a 17-year-old African American male with Stage IV metastatic colon cancer. He was diagnosed five years ago, and had a partial bowel resection at that time and was given a course of chemotherapy. Since colon cancer is very rare in pediatric patients, BK's oncologist typically treated adult patients. One year following completion of chemotherapy, his scans were clear and his oncologist discussed the importance of regular follow up testing. BK's parents were overjoyed by the good news that their son had been "cured," and although they greatly admired and respected this physician, they believed that God had healed their son.

BK felt fine for several years. He was a quiet teenager who generally preferred the company of his family and who spent Wednesday nights and Sundays involved with his church and their youth group. He did not follow up with his oncologist. His parents did not remind him of the conversation with the oncologist, and because BK appeared healthy and they believed him to be cured, they did not heed the doctor's advice. When he turned 16, BK began to experience symptoms, but he kept them to himself. A favorite family story was about how miraculously BK had been healed, and he did not wish to ruin its promise by tempting fate with further testing. Bowel habits are a private matter, so it was easy for BK to conceal the fact that he was often constipated, had dark, tarry stools, and frequently experienced abdominal pain and bloating.

One morning, BK was in so much pain that it was impossible to conceal his discomfort from his mother. She took him to the Emergency Department of the hospital where he had been treated previously. He was diagnosed with a significant bowel obstruction. Surgery for the obstruction revealed metastases to his lungs, liver and abdomen, and he was again started on chemotherapy post-operatively. He was unable to tolerate anything by mouth and had a feeding tube placed. Already tall and thin at this point, BK appeared severely cachectic and dehydrated.

BK's mother stayed at his bedside nearly 24/7, and his father and other family members came and went frequently. BK's mother was very reserved, and although she was pleasant to the nurses and other health care providers who came to her

© Springer International Publishing AG 2017
L. A. Roscoe and D. P. Schenck, *Communication and Bioethics at the End of Life*,
https://doi.org/10.1007/978-3-319-70920-8_7

son's room, it was clear that she preferred privacy. She had a strong Christian religious faith and tremendous trust in the adult oncology physician who had been treating her son since his initial diagnosis. Over time, one of the chaplains befriended BK's mother, and they prayed together. The chaplain also discovered that BK's mother loved scripture, particularly Bible verses that spoke of miraculous cures and the healing power of faith. The chaplain became close to BK as well, who seemed to appreciate having someone in the hospital able to communicate with his mother and offer her some comfort. BK himself appeared to need comfort too; the muscles in his neck would tense up when he thought no one was looking. He deferred to his mother, however, when he was asked about pain levels, leaving her to claim that he did not want any narcotics or any pain medication that would interfere with his ability to communicate with his family and with God. BK acknowledged having pain but would then decline any adjustment that would increase his pain medication and decrease his level of alertness.

BK's mother shared with the chaplain that her son had been fighting cancer for the past five years and was not ready to "give up." She said she had given him permission to "go home to God" if God called him. She added she felt relieved that BK was continuing to receive chemotherapy because that honored his desire to fight his disease and to live. She expressed the belief that since BK had been "cured" of his cancer before, it was likely that would be the case again. She also said she did not want to prolong her son's suffering, and that she was certain God's plan would prevail.

The hospitalist assigned to the case became concerned about inadequate pain control and requested a palliative care consult, but he neglected to tell BK or his mother and they were completely unprepared. They were frightened and upset to have a new doctor come on board unannounced, especially one who was pressing for more pain medication. The palliative care physician was an expert in pediatric palliative care, but he soon became aware that any mention of BK's prognosis, changes to his medication, or any discussion about stopping the current round of chemotherapy was most unwelcome. He asked BK's mother about her discomfort with having her son's pain medication increased, and she replied, "We certainly don't want our son to be a drug addict!" The doctor explained, "Addiction is not really a concern given your son's advanced disease, fears about addiction are not a realistic concern at this point." BK's mother looked up from the Bible on her lap and said to the physician in a quiet but firm voice, "Please leave now. And don't come back."

The social worker assigned to the palliative care team visited with BK during a rare morning when his mother was not there. She talked to him about advance directives, and he said he would like to have one. They worked on it together, and BK recorded his wishes for continued aggressive care. The social worker asked him about his pain level and what he understood about his disease. BK said, "I hurt a lot a lot of the time. I know I'm close to dying and I'm okay with that, but I can't talk about either of those things because that would cause my mom pain and I couldn't stand that." On the way out of BK's room, the social worker spoke to two nurses who had been taking care of him. Both said they were upset about BK's increasing levels of pain and discomfort, as well as by his mother's unwillingness to allow her son to have any relief. The social worker left BK's room and called for an ethics committee consult. By now BK had been on the in-patient oncology unit for 2½ months, and he was clearly declining.

The ethics consult reflected the high levels of moral distress experienced by the nurses who were required to provide aggressive care for BK rather than the pain control and symptom management they felt was more appropriate given his terminal prognosis and deteriorating condition. All team members felt they were doing harm by continuing to provide the aggressive treatment demanded by BK's mother. Even though BK's advance directive confirmed his wishes to continue aggressive treatment, the nurses on the unit doubted that was a true expression of his treatment preferences. The nursing staff also felt that poor care coordination had resulted in BK's mother receiving conflicting information about her son's condition. The trusted family oncologist said they should "stay the course;" the hospitalist and palliative care team members, as well as the nurses, believed further aggressive treatment to be futile and unnecessarily burdensome, but had been politely "excused" from BK's room any time they attempted to have a conversation about changing the goals of care. The nurses caring for BK felt that he did not have a voice in his care since he continually deferred to his mother, even when asked about his pain levels. The nurses also believed that the oncologist BK's mother trusted implicitly had given her unrealistic expectations about her son's prognosis, and that her belief in miracles was making it impossible for her to make decisions in her son's best interests and that realistically reflected his prognosis and need for better pain management. Although he was not legally an adult, BK was a young man who had dealt with serious illness for the past several years, and the health care professionals caring for him felt he should have at least some influence over his care. The oncologist agreed BK should participate in treatment decision-making, and he pointed to the advance directive as evidence of his agreement with the plan of care.

The ethics committee recommended that BK and his family consider discontinuing aggressive treatment and adopting a palliative plan of care. The committee felt that continued chemotherapy was futile and was adding to BK's distress. They also recommended that dialysis, ventilator support and cardiopulmonary resuscitation would be inappropriate, should events occur for which these procedures would otherwise be indicated. The palliative care social worker visited with BK and his mother and told them what the ethics committee had recommended. In her quietly commanding way, BK's mother said, as she had to many others, "please leave, and tell everyone else who has been colluding behind my back to make these recommendations that they are no longer allowed to care for my son." Only the adult oncologist who had "cured" BK in the past would be allowed to come into his room. The palliative care team, and even the chaplain who had previously enjoyed a warm relationship with BK and his mother, were banned from further contact.

BK was discharged to home for a few days during Christmas, but he returned to the hospital shortly after when his condition again deteriorated. Soon after readmission he developed fevers and hypotension. He was transferred from the oncology unit to the Pediatric Intensive Care Unit (PICU) where he developed septic shock. He was put on a ventilator and was moved back to the oncology unit when his condition stabilized. As before, BK's mother refused to allow her son's pain medication to be increased to levels even her trusted oncologist felt were warranted. BK died three weeks later, approximately 3 months before his 18th birthday. His death was painful and private. His family kept the door closed and clearly indicated that they were handling BK's situation with God; no one else was invited.

Discussion Questions

1. Under what circumstances can a patient and/or family member effectively bar medical professionals from doing their jobs and providing appropriate patient care?
2. What role should seriously ill children have in specifying, or agreeing to, plans of care, especially as they approach the legal age of adulthood?

A Bioethicist Responds

By any measure, this is an extraordinarily unfortunate case. Colon cancer is rare in pediatric patients. Overall, colon and rectal cancer in the U.S. ranks fourth behind cancers of the breast, lung and prostate, and as such it accounts for only 8% of the total of all cancers. The average age at time of diagnosis is in the mid to late 60s for all races, and both sexes; it is most frequently found among persons aged 65–74 (24.0% of all those diagnosed). Specific data reveal the highest incidence to occur among Black males, where 59.2 new cases are reported per 100,000 persons, age-adjusted, by race/ethnicity and sex. The disease is so rarely seen in persons under 20, regardless of age, race or sex, that the figure for colon and rectal cancer for all persons, of all races and both sexes under the age of 20 represents only .01% of the entire total of these cancers for all persons, all races and ages, and both sexes (Howlader et al. 2016). In fact, during the time period when BK became ill, and up through the time of his death, there were no cases of colon or rectal cancer reported to the SEER Registry for persons in his age group. BK was most unlucky, indeed.

Cancer presents a challenge to a patient of any age, particularly one as young as BK, but it was not as though he did not also have a few things in his favor: One year post-treatment his scans showed no evidence of disease; his parents believed him to have been healed by God, which must have been very supportive to him at the time; and, for several years he apparently enjoyed the life of a normal, quiet teenager. One could even imagine BK growing into his own person during these years, as would most teens at this stage of their lives, especially if he, too, believed fervently in the healing granted to him by God. Nonetheless, the subsequent onset of symptoms, coupled with his embarrassment over bowel issues and the desire to protect the favorite family story of his miracle cure from cancer, all provided for the crucial turning point in this case, the point at which BK realized he had lost control over his life and the recognition that someone else was in control. Perhaps BK might only have been subconsciously aware of the shift in locus of control at that point, but it is hard to image he could have remained oblivious to it for very long.

The medical facts of this case, together with the increasing pain levels and discomfort acknowledged by BK and those caring for him, leave little doubt as to the suffering created by this disease; it was his mother who demonstrated an attitude of singular nonchalance, if not to say callous indifference, to his pain and discomfort, leaving her to stand in stark contrast to virtually everyone else. Thus, by the time BK had been on the in-patient oncology unit for 2½ months and was clearly declining, and a chaplain, a hospitalist and the palliative care team with its own social worker had all been marshaled to help, a number of professionals were prepared to advocate for BK's welfare, to ensure his good. Unfortunately, their collective view of what was in his best interest was not at all consonant with that of his mother.

There are several ethical issues to be addressed here. The one that may be most troublesome has to do with the unresolvable friction among the multiple players involved in BK's care: his mother, the chaplain, the hospitalist, the palliative care team, the social worker from pediatric palliative care, the nurses, and the outside oncologist trusted by his mother. It seems fairly clear in reading the case narrative that BK's mother rather quickly developed an "us-vs-them" perspective, where she felt compelled to protect her son from unwanted intrusions into what she was convinced would be God's plan for a cure and where she feared BK would become too sedated or addicted to narcotics. Perhaps this conflict should more accurately be described as "me-vs-the rest of you," inasmuch as BK's mother presumed to speak wholly for her son regarding any medications or treatment whatsoever. She maintained this stance to the moment of his death, finally sidelining her heretofore trusted oncologist who had eventually recommended increasing pain medications during BK's final hospitalization. Situations such as these are absolute nightmares for everyone involved: families, physicians, nurses, chaplains, social workers, therapists of all types, aides and, most of all (unless they may be totally incapacitated), the patients themselves. It is difficult to say what, if anything, might have been done differently in order to provide for a better outcome for BK, to have eased his dying process. This albeit loving mother ultimately refused to communicate whatsoever with the highly skilled, experienced and very compassionate professionals caring for her son. Moreover, she was the legitimate surrogate for her minor child who thus held his autonomous rights in her own hands.

This is not a unique story, yet research on the subject of successful communication between families of pediatric patients and physicians and/or nurses is relatively limited. An interesting study by Durall, Zurakowski and Wolfe about advance-care discussions (ACDs) for children with life-threatening conditions revealed that clinicians perceived the most common barriers to be unrealistic parental expectations, differences between clinician and patient/parent understanding of diagnosis, and lack of parent readiness to have the discussion (2012). Less than one-third of clinicians believed that ACDs typically happened at the right time during the course of the patient's illness, and more than 90% of them responded that discussions of overall goals of care should occur either at time of diagnosis or during a time of stability. Nonetheless, the majority of both physicians and nurses reported that these types of discussions happen much later in the patient's illness.

Clearly, a host of factors may figure in the overall issue of communication between parties in a health care setting, perhaps none more significantly than cultural and religious ones. In view of the increasingly multicultural nature of society in the United States, and in view of the increasingly difficult task facing healthcare providers who must try to offer appropriate care for persons with different life experiences, beliefs, value systems, religions, languages and notions of healthcare, Wiener and her colleagues set out to explore and review how culture and religion inform and shape pediatric palliative care (Weiner et al. 2013). They found seven distinct themes: the role of culture in decision making, faith and the involvement of clergy, communication (spoken and unspoken language), communicating to children about death (truth telling), the meaning of pain and suffering, the meaning of death and dying, and location of end-of-life care.

One of the domains identified by Weiner and her group is that of the meaning of illness, dying and death, and they note that such meaning is clearly not static across cultures. Himelstein and his colleagues offered a recommendation that would seem applicable to various cultures: they argued for including a spiritual assessment in the pediatric palliative care plan, an assessment developed from a review of issues such as the child's hopes, dreams, meaning of life, views on prayer and ritual, and beliefs regarding death (Himelstein et al. 2004). It would at least seem to be applicable to various cultures provided the person or persons conducting the spiritual assessment had in-depth knowledge of the patient's culture, were fluent in her/his language, were skilled in conducting such assessments, and were trained in pertinent issues related to cultural competency.[1] In the case of BK, there appeared on the surface to be no serious impediments to the satisfaction of these requirements, and thus to the effective completion of a spiritual assessment. However, there was no way to use whatever spiritual assessment might have been available as part of an implemented care plan because of BK's mother's lack of receptiveness.

There was no meaningful care plan for BK at all. Those responsible for his care in the hospital could not all be expected to function together smoothly as a unified team given the resistance put up by his mother. BK's mother was determined to remain in rigid control of her son, his room, his caregivers and the entire environment in which he had to experience the final chapter of his life. In so doing, she was not only in total control of BK and everything surrounding him, but she effectively usurped the autonomy of this young man despite the fact that he was of capacity and had reached an age where he was capable of speaking for himself, certainly with regard to how he was feeling and as to whether or not he would like more or less pain relief. The result is that BK suffered unconscionably and unnecessarily on more than one level: physically, spiritually and mentally.

One more issue might bear consideration here. Blinderman presents a detailed ethical analysis of the case of a 65-year-old woman dying of cancer and complaining of severe pain, but whose family surrogates wished to minimize her sedation and confusion and increase her alertness and ability to communicate with them (2012). While the patient in this case was a mature adult who lapsed in and out of capacity, and while BK was a minor, family surrogates represented both patients. The argument Blinderman crafts in developing an answer to the question of whether surrogates have a right to refuse pain medication for incapacitated patients is a very compelling one. A full review of how he constructs that argument cannot be undertaken here; suffice it say that he establishes the position that "suffering is bad and that we have an obligation to prevent it or reduce it when it exists" (2012, 302). He also points out that a legal basis for the right to pain relief in the U.S. can be found in the Supreme Court case of Vacco v. Quill (1997); and, in extrapolating a bit from the New Jersey Supreme Court's *In re Conroy* decision (1983), he points out that the best interest standard simply asks: What would a reasonable person in the patient's circumstances consider to be the balance between the benefits and burdens of a particular treatment? Blinderman goes on to state: "If a given treatment (e.g., morphine) is believed to decrease the burdens of life (e.g., terminal cancer pain) and is of benefit (e.g., decreases pain severity) a reasonable person in such circumstances would not refuse such a treatment, even if the treatment were associated with some degree of sedation. Surrogates who attempt to refuse strong opioids that are effective in relieving pain are not acting in the best interests of the patient and are, therefore, not meeting the minimum standard for surrogate decision making" (303).

Happily, in Blinderman's actual case, a resolution was achieved through multiple conversations with the patient's family, and he concluded his article with a statement of his position that "end-of-life medical decisions, including the palliation of pain and other symptoms, should be made together with the patient, or when the patient lacks capacity, with the patient's legally appointed surrogate decision maker or family, to ensure that the patient's specific goals and values are upheld (303)." In the end, Blinderman suggested a very balanced approach that is designed to protect both the unique values and goals of patients and their expressed wishes or best interests, as well as their health care providers' moral obligation to treat pain and suffering, particularly when it concerns vulnerable incapacitated and terminally ill patients (303). He also made it clear that once a treatment plan has been established for an incapacitated patient, and should that plan require opioids, the surrogate or family member should not alter the plan, the sole exception to which should be when the patient had given prior instructions to the surrogate or family member to refuse pain medications on her behalf should she become unable to do so on her own.

Now, it might be argued that Blinderman's piece has little to offer with regard to BK's situation. Blinderman's patient was a mature adult who became incapacitated, and as he reports, the palliative care team was finally able to resolve the issues there; BK was a minor whose surrogate ends up categorically refusing to communicate with anyone. What seems instructive nonetheless is Blinderman's final sentence: "Hospital ethics committees, together with pain palliative care specialists, should draft policies supporting the physician's obligations to treat pain and other symptoms in incapacitated, terminally ill patients over family objections (304)."

This suggestion by Blinderman might well appeal to many physicians and bioethicists, yet hospital administrators (and especially their risk managers) are far more likely to be chary of it. But let us suppose for a moment, just for the sake of argument, that the institution in which the case of BK occurred had such a policy in place. Could such a policy have been used to support the palliative care physician's appeal to wrest control of care from the mother in order to treat BK's pain and discomfort? An interesting question, but even posing it assumes that everyone in this unfortunate case, including the institution itself, erred in not forcing the issue in some way. And to suggest that would be most unfair. One has to consider that BK himself insisted on following his mother's lead even though he also admitted to being in great pain and would have preferred more pain relief; and, it is quite understandable how things develop over time in these situations and how those involved may become reluctant to risk law suits when it may not be clear that charges of "child abuse" (i.e., on the part of BK's mother) could be firmly established. Nonetheless, if a balanced view is to be attempted here, it would not be unfair to say that this is a case where the obligation to treat pain and discomfort in a dying, extremely vulnerable patient was not observed.

A Health Communication Scholar Responds

BK is the central figure of concern is this case, and perhaps the one that evades the closest scrutiny. We know he has been sick for a third of his life with a life-threatening illness that is very rare in teenagers and young adults. It is difficult to be a seriously ill child, and being sick with an "adult" disease is even more distressing. Age is the number one risk factor for colorectal cancer; more than 90% of people diagnosed with the disease are 50 or older and the average age at diagnosis is in the mid to late 60s.[2] In general, teenagers are not likely to get colon cancer. But some genetic conditions and inherited mutations do predispose younger people, including teens and younger children, to developing colorectal cancer.

People with Familial Adenomatous Polyposis Syndrome (FAP), a genetic disorder, have a nearly certain chance of having colorectal cancer, usually diagnosed by age 45, if screening and treatment are not undertaken. This syndrome causes the body to create thousands of adenomatous polyps in the colon, starting in adolescence. One of 10,000 babies will be born with FAP. If a person of a first-degree relative has been diagnosed with FAP, or if there is a family history of the syndrome, colon cancer screening should start between the ages of 10 and 12, with an annual flexible sigmoidoscopy to follow. Bowel surgery to remove the colon (colectomy) is the foremost treatment modality for FAP. People with Hereditary Nonpolyposis Colorectal Cancer Syndrome (HNPCC), also known as Lynch syndrome, also have a nearly 80% of developing colon cancer, and increased risks of developing uterine, stomach, bile duct and urinary tract cancers. People with HNPCC may also have aggressive tumors, which can grow and spread faster than average. People with HNPCC should begin screening by the time they are 20 to 25. People with a first-degree relative who has been diagnosed with colon cancer should begin screening 10 years prior to the age at which the family member was diagnosed (or at 20–25 years old, whichever comes first). A colonoscopy every one to two years may follow the initial screening. We do not know anything about BK's genetic disposition to colon cancer, or whether there was a family history of the disease. We only know that he contracted it at a very young age.[3] The oncologist who treated BK was an expert in colorectal cancer in adult patients, and no doubt he did the best he could to treat BK effectively and compassionately with the tools and treatments he had available. We have no information, however, regarding his experience in treating pediatric oncology patients, or whether he thoroughly inquired about BK's family medical history.

There are many difficulties that occur when a hospitalized patient transfers between pediatric and adult hospital floors and services and health care professionals who normally treat children and those reserved for adult patients. The expected difficulties around coordination of care are exacerbated in such circumstances, and there is scarce research literature on best practices to guide policy changes (LoCasel-Crouch and Johnson 2005). In BK's case, his mother trusted the oncologist who treated BK upon his initial diagnosis and who was involved in his care five years later. Although likely highly qualified, this physician was not an expert in dealing with parents of seriously ill children, of managing pain in seriously ill children, or in coordinating the care that pediatric patients and their families might require (including Child Life Specialists, etc.). Eventually it appeared that all the right players were involved—palliative care, ethics, various other specialists and intensivists—but it is not sufficient just to have all the right people involved asynchronously. They must be on the same page in terms of the information given to the patient and family, and they must be committed to developing and implementing a plan of care that respects their various disciplinary contributions informed by the current state of medical science, the values and preferences of the patient and family, and the medical facts of the patient's condition. This is best achieved in a face-to-face meeting, which may be difficult to arrange (but should not be impossible).

BK's mother may not have had all the information she needed in order to make decisions that were in her son's best interest. First and foremost, she did not acknowledge or know what BK himself was experiencing, knew, or desired—the first key step in honoring a person's autonomy is knowing what they want for themselves. In BK's case, he seemed consistently to want what would cause his mother the least pain and distress, including spiritual distress. It is not difficult to imagine that a young man who has been seriously ill for much of his life would look to his parents for guidance, and he may well have lacked the emotional maturity to make or even to formulate independent ideas about what he valued and cared about. There also was a lack of communication between the palliative care team, the hospitalist, and the oncologist in terms of what they were telling BK and his mother and what they were sharing amongst themselves. It is quite possible that BK's mother really did believe her son had been cured earlier, and that she steadfastly believed another miraculous cure was likely or at least possible. Again, since we do not know BK's family health history, we can only guess at his parents' level of health literacy and ability to interpret and act upon the information she was given about BK's condition and prognosis (Berkman et al. 2011).

Physicians and other medical professionals are likely to roll their eyes at family members who insist that God will make the decisions and that a miracle is sure to happen. Just because BK's mother was a woman of strong faith who was hoping and praying for a miraculous cure for her son does not mean she did not believe nor understand the medical information she was given about his prognosis. Recent research has shown that parents of seriously ill children use both medical and spiritual resources when making decisions about their children's health care (Davidson 2016). For parents of seriously ill children, often every choice they could make appears bad. For example, consenting to surgery may be distressing and painful to their child and/or of uncertain benefit, but not consenting to surgery also brings anguish and uncertainty. Sometimes asking or hoping for a miracle is more a reflection of a need for more time, and a need for a spiritual sign, rather than a denial of the seriousness of their child's condition or prognosis. In BK's case, however, we do not know to what extent his mother understood and accepted her son's dire prognosis. Her unwillingness to bring him for follow up testing after his earlier remission might indicate either a deep denial or a serious misunderstanding of her son's chronic medical condition.

It might also be the case that BK's mother did not want additional medical information, and that may explain why she sent away the palliative care team who tried to challenge the plan of care. The Theory of Motivated Information Management (TMIM) posits that the likelihood we will seek information depends on our perceived need for it, our coping ability, and the way in which the information is conveyed (Afifi and Weiner 2004). It focuses on interpersonal sources of information, and it predicts that when the stakes are high, most people prefer face-to-face communication from highly trusted sources. It appears that BK's mother felt she had all the information she needed from her trusted sources, namely, her son's oncologist, and God.

The last issue to be addressed concerns the level of pain management that BK apparently needed but was not willing to access. Many patients prefer some level of discomfort along with the ability to interact with family and friends rather than complete pain control that renders them semi-conscious. For spiritually inclined families such as BK's, the ability to pray and interact with God is also an important value. The risk of becoming addicted to narcotic pain medication is real, but limited by several factors, and overall the risk is quite low. Fishbain and colleagues (2008) conducted a structured evidence-based review of all available studies on the development of abuse/addiction and aberrant drug-related behaviors in chronic pain patients with nonmalignant pain who were exposed to chronic opioid analgesic therapy. Their results revealed that for patients with no previous or current history of abuse or addiction, the risk of becoming addicted or of abusing opioids was less than 1.0%. In BK's case, it was extremely unlikely that he would become addicted before he died, but because his mother was not willing to concede that his life expectancy was foreshortened the risk of addiction was real to her. Perhaps a family history of addiction was another pertinent medical clue that this very private family chose not to share.

BK died in pain, after being subjected to months of life-prolonging medical treatment that did nothing to maximize his chances of recovery or comfort. His death was distressing to the medical professionals caring for him, and it is likely also distressing to those reading this case study. It is important to remember that even so-called "good" deaths involve some level of compromise and distress, and perhaps the most painful thing for BK would have been to witness the pain and suffering his death caused his parents and family.

Notes

[1]Cultural competency is a huge field. Even with delimiters set to healthcare, a literature search will yield over 2600 items; by comparison, relatively little research has been done in pediatric and hospice, or end-of-life care. The following may be useful in developing a general overview of this field:

Brunger, Fern. 2016. Guidelines for teaching cross-cultural clinical ethics: Critiquing ideology and confronting power in the service of a principles-based pedagogy. *Journal of Bioethical Inquiry* 13: 117–132.
Cheng, Tina A., Mickey A. Emmanuel, Daniel J. Levy, and Renee K. Jenkins. 2016. Child health disparities: What can a clinician do? *Pediatrics* 136: 961–968.
Houghson, Jo-anne, Robyn Woodward-Kron, Anna Parker, John Hajek, Agnese Bresin, Ute Knoch, Tuong Phan, and David Story. 2016. A review of approaches to improve participation of culturally and linguistically diverse populations in clinical trials. *Trials* 26: 263.
Purnell, Larry. 2016. Are we really measuring cultural competence? *Nursing Science Quarterly* 29: 124–127.

[2]*Detailed Guide: Colon and Rectum Cancer: What Are the Risk Factors for Colorectal Cancer?* American Cancer Society. 7 Mar. 2006. 23 Jun. 2006 [http://www.cancer.org/docroot/CRI/content/CRI_2_4_2X_What_are_the_risk_factors_for_colon_and_rectum_cancer.asp].

[3]There are some other rare genetic mutations that might cause colon cancer in young patients. Peutz-Jeghers syndrome is a very rare genetic syndrome that is characterized by polyps in the gastrointestinal tract and may be accompanied by freckles around the mouth, hands and feet. Peutz-Jeghers syndrome greatly increases the chances of developing colon cancer. It is caused by a defect in the STK1 gene and can be diagnosed through genetic testing. Juvenile polyposis is a condition that causes multiplication of the polyps in the gastrointestinal tract of young children. It will lead to colorectal cancer if untreated. Turcot syndrome is another very rare syndrome that can increase the risk of developing adenomatous polyps. There are two variations of Turcot—one that mimics FAP and one that mimics the mutations seen in HNPCC. Turcot syndrome may also increase the risk of developing brain cancer.

References

Afifi, W.A., and J.L. Weiner. 2004. Toward a theory of motivated information management. *Communication Theory* 14: 167–190. https://doi.org/10.1111/j.1468-2885.2004.tb00310.x.

Berkman, Nancy. D., Stacey L. Sheridan, Katrina E. Donahue, David J. Halpern, and Karen Crotty. 2011. Low health literacy and health outcomes: An updated systematic review. *Annals of Internal Medicine* 155: 97–107. https://doi.org/10.7326/0003-4819-155-2-201107190-00005.

Blinderman, C.D. 2012. Do surrogates have the right to refuse pain medications for incompetent patients? *Journal of Pain and Symptom Management* 43: 299–305.

Davidson, Lindy G. 2016. Spiritual frameworks in pediatric palliative care: Understanding parental decision-making. (Unpublished doctoral dissertation). University of South Florida, Tampa, Florida.

Durall, A., D. Zurakowski, and W.J. Joanne. 2012. Barriers to conducting advanced-care discussions for children with life-threatening conditions. *Pediatrics* 129: 979–982.

Fishbain, D.A., B. Cole, J. Lewis, H.L. Rosomoff, and R.S. Rosomoff. 2008. What percentage of nonmalignant pain patients exposed to chronic opioid analgesic therapy develop abuse/addiction and/or aberrant drug-related behaviors? A structured evidence-based review. *Pain Medicine* 9: 444–459.

Himelstein, Bruce P., Jeanne M. Hilden, Ann M. Boldt, and David Weissman. 2004. Pediatric palliative care. *The New England Journal of Medicine* 350: 1752–1762.

Howlader, N., A.M. Noone, M. Krapcho, D. Miller, K. Bishop, S.F. Altekruse, C.L. Kosary, M. Yu, J. Ruhl, Z. Tatalovich, A. Mariotto, D.R. Lewis, H.S. Chen, E.J. Feuer, and K.A. Cronin, eds. 2016. SEER Cancer Statistics Review, 1975–2013, National Cancer Institute. Bethesda, MD, http://seer.cancer.gov/csr/1975_2013/, based on November 2015 SEER data submission, posted to the SEER web site, April 2016.

LoCasel-Crouch, J., and B. Johnson. 2005. Transition from pediatric to adult medical care. *Advances in Chronic Kidney Disease* 12: 412–417.

New Jersey, Superior Court. 1983. Appellate Division. *In re Conroy.* Atl Report 464: 303–315.

Vacco v. Quill 521 U.S. 793 (1997).

Weiner, L., D.G. McConnell, L. Latella, and E. Ludi. 2013. Cultural and religious considerations in pediatric palliative care. *Palliative Support Care* 11: 47–67. https://doi.org/10.1017/S1478951511001027.

Case 8—A Young Woman's Wish to Die

VH was a 19-year-old woman admitted to the Pediatric Intensive Care Unit (PICU) of a large, tertiary care hospital following a T-bone motor vehicle crash.[1] She was alone in the car. Her injuries included a C-2 C-3 fracture[2] that resulted in quadriplegia. VH was ventilator dependent and had been placed in the PICU for lack of beds in the adult ICU at the time of admission. Upon returning to full consciousness, VH was made aware of the severity of her condition and that she may or may not recover any motor function. This news resulted in profound sadness, and she was evaluated by a psychiatrist for clinically significant depression.

VH was Hindu. She and her family had moved to the United States several years prior to her accident and were settled with the help of her father's brother. The patient's parents were not happy living in the United States, however, and they planned to return to India when their children had completed their educations. The patient herself was a student at the local community college in a physical therapy program. Despite her young age, she was looked to by members of her large, extended family—some of whom were not fluent in English—for guidance in negotiating life in the U.S. Her parents operated an independent grocery store as their primary means of support. They were not wealthy, but they supported themselves. VH's uncle, fluent in English, spoke for the extended family and acted as the intermediary between the family and the hospital.

The family indicated they might not wish to continue life support should VH continue to be paralyzed with little hope of improvement. The uncle stated, "In our culture, the family makes decisions in this kind of situation." He explained that according to their cultural beliefs, suffering could be the result of transgressions committed in a past life. He clarified by saying, "VH might now be atoning for transgressions of a past life and must be allowed to be reborn. Her body is only a garment for her soul, and when the body is broken the soul must be freed to go on to the next life." In the family's view, therefore, this reincarnation into a new existence, a new life, may be a much happier one than the one VH currently had.

© Springer International Publishing AG 2017
L. A. Roscoe and D. P. Schenck, *Communication and Bioethics at the End of Life*,
https://doi.org/10.1007/978-3-319-70920-8_8

The hospital Risk Manager explained to the uncle and family that the legal age for giving consent in the United States was 18 and that VH would, therefore, be the only one allowed to consent to withdrawal of treatment if her physicians determined she had decision making capacity. The uncle said they wished to be respectful of this, but that they also felt the issue to be a family matter. He also stressed, however, that the family wished to avoid any media or legal involvement, sensitized as they were to the Schiavo case.[3] The patient's physicians strongly urged that no decision be taken until a tracheotomy had been performed to allow continued ventilator support and the patient's current depression be given time to respond to treatment. They felt that once those objectives had been achieved VH would be better able to assess her situation with the help of seasoned medical professionals and family members, take time for careful reflection, and then make her decision. They also encouraged the family to wait (perhaps some months) until VH's prognosis could more adequately be assessed.

An ethics consult was called. The ethics committee strongly supported the recommendations of the physicians and other health care team members, particularly with regard to soliciting the patient's wishes, if at all possible, noting that a competent patient has a right to refuse treatment. The family agreed to wait so that the recommended treatments could be performed, although they demurred on waiting months for further assessments. The uncle stated that while he understood the patient, and not the family, would need to be the one to give consent to any decisions about discontinuing treatment, he said, "VH is a good girl who will do the right thing," reinforcing this with a proud, loving, avuncular smile. The family also indicated through him that VH had begun communicating with them through eye gestures that she did not want to be a "burden on the family," and that she wished to be allowed to die.

The tracheotomy was performed. VH slowly began to form words, though she could not speak aloud. The medical team read her lips for most of their communication with her. The psychiatrist treating her for depression visited daily, as did the licensed clinical social worker assigned to her case. She was also seen daily by the PICU intensivist to keep her up to date about her medical condition and to continue to assess her understanding of her situation. She was given detailed information about options for rehabilitation and financial resources that might be available to her and her family. Her depression abated, and she appeared to have decision-making capacity. VH indicated understanding of her condition and that if taken off the ventilator she would die. She reaffirmed her wish to be allowed to do so and indicated she was unafraid and at peace with that decision.

The ethics committee met again with the uncle and several of VH's family members. "We respect the ethical, legal, and medical positions of the hospital," said VH's uncle, "but our religion and culture are important to us, as is VH." He also emphasized the apparent wishes of the patient herself to stop life support. The ethics committee, believing that VH was making a fully considered and informed decision, recommended to the physicians that the ventilator be withdrawn. The attending physician discussed this with VH and her family, and he reassured everyone that VH would be sedated before the removal of the ventilator and that she would experience no discomfort. The family asked that this be done between 12:00 noon and 1:30 p.m. in accordance with their religious beliefs; they also asked that they be allowed to practice certain religious rituals in the room privately with VH before the ventilator support was discontinued. These requests were granted. The family and VH said their good-byes, and she expired peacefully at 1:20 p.m. with her uncle and several other family members present.

Discussion Questions

1. How can hospitals and physicians in the U.S., where ethical deliberations most commonly focus on the patient-physician dyad, adequately support patients where more family-centered, communal decision making might be preferred?
2. Would it have been helpful in this case to have a chaplain involved? What role can chaplains play in helping hospital personnel and families think through end-of-life issues where religion plays a significant role? How can chaplains serving in hospitals sponsored by a specific religion best be of service to patients of very different religions and cultures, or patients espousing no religious beliefs at all?

A Bioethicist Responds

Full disclosure is in order here once again. This writer was a member of the ethics committee mentioned in the narrative above, and it has to be said that the ethical issues weighed heavily upon all members of the committee at the time. Yet when this case was later brought up for discussion in a university biomedical ethics seminar, the first-blush reaction of the class, virtually year after year, was generally that there were no ethical issues here. If Professor Schenck asked if they were sure that they saw no ethical issues, the response would go something like, "No, once VH was found to have decision making capacity, which she eventually was, it was merely a matter of allowing her to exercise her autonomy. And, the docs and the hospital did allow her to do that, so what's the issue?" Fine, one might say, but Professor Schenck felt there just might be a few more wrinkles to the story, not the least of which was the interesting intersection of two very different cultures, and there was considerable challenge in attempting to enlarge the optic of some students as to what was happening in that regard. Nonetheless, Professor Schenck himself remained puzzled for years, if not to say uncomfortable, about one particular issue in the case until one semester when a number of male and female students of Hindu background enrolled in the seminar. Before further discussion of that, however, some preliminary observations may be useful.

One of the most obvious features to be noted in a wide-angle view of this case, as mentioned above, is the juxtaposition of two very different cultures, the Hindu cultural values and customs of the patient and her family and those of the western hospital and health care team. One might almost suggest a "clash of cultures" here, but that would be too strong a term since no real struggle ensued and a resolution was found to what appeared to be problematic at first. Fortunately for all concerned, this hospital was not only truly dedicated to serving all persons, regardless of race, religion, color, creed or national origin, but it had taken seriously the provisions of the law established in 2000 as the *National Standards for Culturally and Linguistically Appropriate Services in Health Care* (commonly known as CLAS Standards).[4] The physicians and the entire staff were, therefore, fully committed to making every effort to achieve an understanding of VH's feelings, beliefs and wishes, and those of her family, as well as to finding ways to communicate the concerns of the health care team for her and her family, and to communicating the values and ethics of the hospital and the medical team in ways that VH and her family could understand.

Some of the research data reported in recent years on people from the Indian subcontinent regarding their attitudes and beliefs at the end of life might seem surprising to people in the west. Loiselle and Sterling reported on a grassroots cancer care hospice in Bangalore, India, pointing out how the significant rise in late-stage illness diagnoses, such as cancer, is placing enormous demands on palliative care (2012). The primary purpose of their study was to demonstrate the need to develop supportive mechanisms for hospice workers in order to ensure an optimal workplace climate, but it also revealed the increasingly complex cultural, socio-economic and religious contexts of the populations these hospice care workers serve. Briefly put, those in the west should not simply assume that end-of-life issues for persons from the Indian subcontinent will naturally embrace Hindu beliefs; the overall population exceeding one billion persons is far too complex.

The situation for Asian Indians living in the United States could be even more difficult for physicians and others in health care here to understand. Rashmi Gupta's article on Asian Indian American Hindu cultural views related to death and dying presents the results of a qualitative study designed to fill a gap in empirical research on cultural beliefs and meanings of death and pre- and post-death practices in three generations of Hindu Americans in the U.S. (2011). The study revealed the importance for the dying and the family to say their good-byes; the contradictory attitudes regarding expressions of grief that may be found among Hindus; the fact that Hindu culture is group-oriented or collectivist in nature; and Hindu beliefs about the meaning of death as a transition to another life that helps people feel less anxiety about death than would normally be felt by people of many western cultures. Gupta then calls for those in health care in the United States to develop greater cultural awareness of Asian Indian Hindu beliefs and rituals pre- and post-death, as well as a clear recognition that not all Asian Indian American Hindus will respond alike to end-of-life situations.

Somewhat similarly, Sharma and colleagues (2011) reported from their study that first- and second-generation Asian Indians place high value on close parent-child relationships and feel a strong sense of duty to the family. And, they recommended in fact that "When eliciting Asian Indian patient preferences for end-of-life care, clinicians should consider explicitly asking about preferences related to family involvement in care, decision control, and communication; and explore the role of traditional expectations and specific social realities for each patient. By assessing each of these topics and respecting the patient's choice for individual or family-centered decision making, clinicians caring for Asian Indian patients can maintain respect for patient autonomy while also respecting the patient's cultural values" (2011, 316). With the foregoing in mind, it is time to return to a more detailed view of VH's situation and what appeared, at least at one point, to be the troublesome issue of informed consent.

Recalling the case narrative, VH's uncle had said he understood his niece would have to be the one to give consent to any procedures or withdrawal of treatment, but he also knew her to be a "good girl who would do the right thing." One can well imagine the reactions of the members of the ethics committee upon hearing this statement. While no one said anything in direct response to the uncle at the time, committee members expressed their fears and concerns openly amongst themselves once the uncle and other members of the family had left the room. There was some minimal understanding of the nature of *karma*, atonement and reincarnation, but there was also fear that VH might be subject to family coercion out of a sense of obligation to the family, thereby making the issue of informed consent highly problematic. As reported above, however, things eventually worked out to the satisfaction of both parties and VH was made comfortable while she was removed from life support and allowed to die as she had requested.

This leaves the issue temporarily suspended above, the one that had continued to trouble the present writer for some time despite his agreement with the decision to withdraw the ventilator, albeit sadly, and despite his conviction that this was the right decision for VH and an acceptable decision for her American care givers. What still seemed troublesome, nonetheless, was the idea that VH was of majority age, that she apparently played some sort of leadership role in her extended family, but that "the family would make decisions in this kind of situation." It was only when this confusion was expressed years later in class that an 18-year-old Indian-American woman said, "I think I can help you, Dr. Schenck. You see, in my culture, and even though I've spent most of my life in the United States, I won't be considered an adult until sometime in my 20s. And, there's no fixed age when one becomes an adult. I might be considered an adult by my family when I'm 22, or maybe not until I'm 27; it all depends upon when they, as well as I, feel I've really matured. And, in fact, although other people my age, that is 18, are legally adults in the U.S., I'm still a child as far as my family is concerned in Indian culture. So you shouldn't really think that VH's uncle or other family members coerced her into saying she wanted the ventilator withdrawn or that she wanted to die when she really didn't. When her uncle said that 'she was a good girl who would do the right thing' he meant she would do what she really believed was the right thing not only because of her faith but also because the family knew this to be the right thing, and she would have known as well that this was the culturally, ethically appropriate thing to do. After all, serious decisions such as these are made by the family as a whole, not by one individual alone." This eloquent explanation was given by an extraordinarily bright young woman of Indian origin but raised entirely in America, and now in medical school.

And with that, Professor Schenck looked around the room to see heads nodding, several with knowing smiles on their faces, males and females alike, all of them

Asian Indian American Hindus who had been largely educated in the United States, all confirming what their classmate had said. It finally made sense, and the resolution to the case of VH seemed all the more striking, given the potential for problems and misunderstanding.

A Health Communication Scholar Responds

VH's accident, injury, and death were tragedies for her and her family. Some things worked well in this case nonetheless which prevented this sad story from an even more calamitous ending. It can be hard to make sense of a young person dying well before their time, but VH and her family used their religious and cultural resources to create meaning and make decisions that were consistent with their values and beliefs. Ironically, one of the factors that led to VH's peaceful death and good communication between family members and medical staff members was also a troubling one—her uncle acting as the family spokesperson. Having a single point of contact helped this family cope and make decisions, and it undoubtedly helped the medical team to coordinate VH's care, but we are left to wonder how much of a role VH herself played in this sequence of events. Our ethical principles and communication frameworks in medicine are predicated on the assumption that the unit of interest is the patient-physician dyad. Yet we know in practice that this is rarely the case, especially in intensive care units, where teams of hospitalists, physician specialists, nurses, occupational and physical therapists, respiratory technicians, social workers, chaplains and others all play crucial roles. VH was lucky in that in addition to an apparent abundance of good medical care, she also had the loving support of her close family. But how do we know, as we must ascertain given our focus on the patient-physician dyad, that this patient had an opportunity to speak with her attending physician and express her own preferences?

We cannot look inside this patient's psyche and parse how much of her decision to remove herself from life support was based on personal preference, how much was influenced by family pressure, or how much her physicians swayed her opinion one way or another. We do not know much about VH herself, other than her role as navigator for her family members new to the U.S., and her designation as a "good girl." Is this how she saw herself? She was only 19 years old—was she a mature and independent young woman, or still dependent on the ideas and opinions of her family? Did she adhere devoutly to Hindu religious teachings and beliefs, or was she more secular in the beliefs and values that guided her throughout her short life? Does it matter?

VH's family members believed that their religious and cultural perspective led them to conclude that VH should be allowed to let go of her damaged body and await a future reincarnation.[5] Patients who are judged competent are allowed to refuse medical treatment, including life support, under any circumstances. Given VH's young age and high level of family support, we might be tempted, as were some of the doctors involved in her care, to attempt to persuade her and her family to wait longer to decide whether to discontinue life support. We certainly do not want a severely depressed individual to end life supportive treatment before they could make this decision with a clearer head. We do not want to encourage newly injured patients to refuse life support while in the middle of processing the initial shock and despair over their condition. However, imposing an arbitrary waiting or adjustment period before making this decision also imposes challenges.

There is no ethical or legal difference between withholding or never starting life support and withdrawing life support after a trial period. In clinical practice the decision to withdraw life support is often more difficult. Consider the famous case of Donald "Dax" Cowart.[6] Dax was severely injured in a propane gas explosion that killed his father and left him blind and with third-degree burns over much of his body. He attempted to refuse the excruciatingly painful treatments he received, but his mother consented on his behalf despite his protestations and the fact that his some of physicians believed he was competent to make his own decisions. Nearly a year into his treatment, one of Dax's physicians again refused to let him stop treatment, telling him, "you've come this far, why stop now?" This same surgeon convinced Dax to have "just one more surgery" to restore some of the lost function in his hands, and told Dax that once that was done, he would be able to kill himself if he wanted to without a doctor's help. Dax did try to kill himself at least twice—once by trying to step in front of a truck speeding by along the highway at night, and once by taking an intentional overdose of sleeping pills. Neither suicide attempt was successful. Dax has by now been a successful attorney for many years, and describes his life as happy. He still maintains that he should have been allowed to refuse treatment and die from his injuries. He remains personally and professionally committed to the issue of patients' rights.

VH's case reminded me of some interesting parallels to my grandmother's experiences at the end of her life. My grandmother was a devout Catholic her entire life. She never got a driver's license, and after her husband died in his late 60s, she walked to Mass nearly every day, in every season, for nearly 30 years. About a year before her 90th birthday, she was diagnosed with peripheral artery disease and was in danger of losing two of her toes, which distressed her greatly. She was persuaded to undergo femoral-tibial bypass surgery (also known as infra-popliteal reconstruction) to bypass the diseased blood vessels in her lower leg or foot.[7] My grandmother spent several weeks in the hospital and was eventually discharged to a skilled nursing facility. In hindsight, the recovery from amputation of her toes would likely have been easier than the recovery from the bypass surgery. Even though she had the loving support of her two daughters and her extended family, this was still an extremely difficult period of suffering for her. She died a few months after her 90th birthday. At her funeral mass, the priest described her post-surgical suffering and institutionalization as her required time in purgatory. While in line with Catholic teachings about the role of suffering in human life, this explanation of her end-of-life care struck me as cruel and unnecessary.

Religion can be a tremendous resource and a source of meaning, comfort, and strength. Catholicism was a central feature of my grandmother's life, and while the priest's words offered me little consolation, I think my grandmother would have understood his description of her end-of-life journey, and perhaps she would have found succor in his acknowledgement of her suffering. Although we do not know how invested VH was in Hinduism, we can hope that the promise of reincarnation was a source of comfort and hope. It is more difficult to find reassurance, at least to this writer, in the idea that her injury may have resulted from past life transgressions; but perhaps belief in the former requires acceptance of the latter. My grandmother was not what some call a "cafeteria" Catholic, one who picks and chooses only parts of the faith. My grandmother accepted it all, including messages about the necessity of suffering. Perhaps VH was also able to hold both the promise of reincarnation and the sanction of injury for past sins close as she and her family made sense of her situation and grappled with the decision to allow her to die.

Our global and diverse society means that those with varying religious and cultural frames of reference are increasingly part of the fabric of the institutions and norms that make up our American way of life. We need to remain mindful of the ways in which our Judeo-Christian roots have formed and continue to shape our institutional practices, the ways we expect families to behave, our taboos about what we can talk about and how, and the individualism that characterizes our ethical principles and processes. We need to continue to seek ways to talk across our differences and find common ground. At first blush, Hinduism and Catholicism, for example, seem to have little in common. But even a fairly superficial examination reveals that both faith traditions believe in an afterlife (either reincarnation or ascension into heaven), acknowledge suffering as an inherent part of the human condition, recognize the importance of family life, and do not insist that "everything" be done to avoid or forestall death. In VH's case, her family and her medical team were able to reach a respectful compromise. They waited for some time to make decisions, but not too long, consented to some procedures but not to others, and respected VH's autonomy along with her desire to rest it within the bosom of her extended family. Her physicians did not insist on an arbitrary period of adjustment or discharge to a rehabilitation facility. They talked civilly to one another and were able to find a compromise between hospital policy and physician expertise and family culture and religious values. The focus on VH and her well-being, as both a patient and a member of a close family, allowed decisions to be made in a timely way that allowed all to be at peace while still acknowledging the tragic situation at hand.

VH was young, and while that might have meant she had the resilience required to adjust to her new life circumstances as a quadriplegic, it also meant she potentially faced a long lifetime of changed opportunities and expectations. VH made a decision that considered all her options, one that was supported by her extended family and understood by her medical team. Her withdrawal from life support and subsequent death was managed with dignity and acceptance. May she rest in peace until….

Notes

[1]Broadside collisions are described as those in which the side of one vehicle is impacted by the front or rear of another vehicle, forming a "T". In the United States and Canada this collision type is also known as right-angle collision or T-bone

collision. Vehicle damage and occupant injury are likely to be severe, but severity varies based on the part of the vehicle that is struck, safety features present, the speeds of both vehicles, and vehicle weight and construction. In VH's case, her small compact sedan was hit on the driver's side by a much heavier vehicle travelling through a red light at high speed.

[2]The cervical (C) vertebrae are the bones comprising the upper spine or neck. There are seven cervical bones, and if any of these are fractured or severely dislocated, the neck can be described as broken. In some cases, the fractured or dislocated vertebrae can cause compression and damage to the spinal cord resulting in neurological symptoms or paralysis. A fracture of the pars interarticularis on the pedicle of C2 (axis) is often called a "hangman's" fracture. This term was coined by Scheider and colleagues based on the observation that these were the types of fractures seen in hanged criminals; falls and automobile accidents are other common causes. This type of fracture is the most common of all cervical spine fractures and is often caused by extreme hyperextension to the neck when the face forcibly strikes a hard surface. Fractures of the C3 vertebrae, although uncommon, have been linked to a higher mortality rate than other cervical fractures. The phrenic nerve can be damaged in a C3 fracture, resulting in paralysis of the diaphragm, a probable cause of higher mortality.

For more clinical information, see Schneider, R. C., K. E. Livingston, A. J. Cave, A., and G. Hamilton. 1965. "Hangman's" fracture of the cervical spine. *Journal of Neurosurgery* 22: 141–154; and Pull ter Gunne, A., A. Aquarius, and J. Roukema. 2008. Risk factors predicting mortality after blunt traumatic cervical fracture. *Injury,* 39: 1437–1441.

[3]The events of the Schiavo case are likely familiar to most readers of this volume. See Case 9 for a brief review.

[4]See, 65FR80865. (Federal Register/Vol. 65, No. 247/Friday, December 22, 2000/Notices.) The final document consists of fourteen Standards. Interested readers will find terms such as culture, cultural and linguistic competence, culturally and linguistically appropriate services, health care organizations, and patients/consumers defined in the *Preamble*.

[5]Hinduism is a broad-based and doctrinally tolerant religion with hardly any single and exclusive test of orthodoxy; therefore, one must not presume to know what the beliefs and practices of a patient might be from the mere fact of his or her self-identification as Hindu. Hindu medicine is patient-specific, and the ethical course of action is influenced by the individual patient's vocation, age, gender, teleological disposition, and cosmo-temporal placement. (See Sharma, Arvind. 2002). The Hindu tradition: Religious beliefs and healthcare decisions. Part of the "Religious Traditions and Healthcare Decisions" handbook series published by the Park Ridge Center for the Study of Health, Faith, and Ethics.

Joint family systems are prevalent and proxy decision making is eminently acceptable. In particular, women often subordinate their own concerns to that of the family in the pursuit of their *dharma* (or place in the cosmic order). According to S. Radhakrishnan, "The interests of the individual may be subordinated to those of the family; of the family to those of the county; of the county to those of the region; of the region to those of the nation; and of the nation to those of the larger world." (See Radhakrishnan, Sarvepalli. 1993. *The Hindu view of life.* New Delhi: Indus (first published in 1927), p. 64.)

Handicapped individuals could be viewed as working out their bad *karma* through their present condition. This view could, however, easily lead one into blaming their bad *karma* for their condition. One widely acknowledged ethical stance is to shift the focus to ask "What is our *dharma* (duty), given that the other person's *karma* brought the person to his or her current state?" (See Sharma, Arvind. 1999. Karma was fouled: How Hoddle did no wrong to the disabled. *Hinduism Today*, September 1999, 13).

[6]Much has been written about the case of Dax Cowart. See for example Cowart, D., & Burt, R. (1998). Confronting death: Who chooses? Who controls? A dialogue between Dax Cowart and Robert Burt, *The Hastings Center Report, 28*, 14–24; and Hurst, Ashley R., Dea Mahanes, and Mary Faith Marshall. 2014. Dax's case redux: When comes the end of the day? *Narrative Inquiry in Bioethics* 4: 171–177. Video documentaries about Dax's case are also instructive, see for example, "A Right to Die?: The Dax Cowart Case. An Ethical Case Study on CD-Rom," by David Anderson, Robert Cavalier, and Preston Covey (available on Amazon).

[7]Most of the time, surgery is only done in cases of severe peripheral arterial disease (PAD), such as disabling intermittent claudication; open sores (ulcers that won't heal); or serious skin, bone, and tissue problems (gangrene). Bypass surgery redirects blood through a grafted blood vessel to bypass the blood vessel that is damaged. The grafted blood vessel may be a healthy natural vein or artery, or it may be man-made. The methods of bypass surgery vary depending on the size of the affected artery and where it is located.

References

Gupta, R. 2011. Death beliefs and practices from an Asian Indian American Hindu perspective. *Death Studies* 35: 244–266.

Loiselle, C.G., and M.M. Sterling. 2012. Views on death and dying among health care workers in an Indian cancer care hospice: Balancing individual and collective perspectives. *Palliative Medicine* 26: 250–256.

Sharma, Rashmi K., Nidhi Khosla, James A. Tulsky, and Joseph A. Carrese. 2011. Traditional expectations versus U.S. realities: First- and second-generation Asian Indian perspectives on end-of-life care. *Journal of General Internal Medicine* 27: 311–317.

Case 9—When Parents Contest an Adult Child's Advance Directive

AH was a 25-year-old woman who had been admitted to the intensive care unit of a large teaching hospital, despite being under the care of the local hospice agency. She had been diagnosed with end-stage cardiomyopathy (weakening of the heart muscle) and currently had an ejection fraction of 10% (normal range: 55–70%). AH made sure the attending physician knew she had an advance directive that specified her care was to be limited to comfort measures only, with the added provision that she did not want to be on a ventilator or to be maintained on a feeding tube for more than 30 days, which contradicted her preference for comfort measures only. This concession resulted from an apparent disagreement with her mother, who wanted her daughter to agree to life-sustaining treatment indefinitely should it become necessary.

AH lapsed into a comatose state shortly after admission to the hospital, and required ventilator support as well as a feeding tube for artificial nutrition and hydration. Her parents, who were divorced, were informed of her poor prognosis and were reminded of her advance directive which specified time limits on life-sustaining treatment were it to be initiated. AH's father was an uneducated man who wanted to honor his daughter's wishes and remove her from life support after 30 days. AH's mother, who had a history of schizophrenia, wanted her daughter kept alive indefinitely and hired an attorney who had previously represented Terri Schiavo's parents.[1]

AH's condition did not improve during the 30 day trial of life sustaining treatment. On the evening of the 29th day of ventilator and feeding tube support, the attorney who had been hired by AH's mother came to the hospital, placed a note in the medical chart indicating that she had the authority to revoke AH's advance directive, and attached her business card to the chart. (This case predated the use of electronic medical records.) At this point, a physician known for his commitment to honoring patients' advance directives rotated onto the unit. He was not intimidated

© Springer International Publishing AG 2017

L. A. Roscoe and D. P. Schenck, *Communication and Bioethics at the End of Life*,

https://doi.org/10.1007/978-3-319-70920-8_9

by the attorney's note, despite her involvement with the infamous Schiavo case, and urged the hospital administration to go to court if necessary to ensure his patient's wishes for discontinuation of treatment after 30 days.

A family meeting was held with the attending physician, nurses and social workers on the unit, ethics committee members, and AH's parents. The attending physician strongly endorsed allowing AH to be removed from life support according to the provisions of her advance directive, which was supported by all medical and ethics committee staff members, as well as by AH's father. Her mother disagreed and threatened to bring legal action against the hospital if life support measures were discontinued.

AH's mother and her attorney filed suit against the hospital. During the subsequent court hearing, the mother's mental instability was notable. She argued that the signature on her daughter's advance directive had been forged, even though she also conceded that it matched her daughter's recent signatures on consent forms signed upon her hospital admission. AH's mother also produced a note on scraps of paper taped together that stated her daughter wanted her advance directive rescinded, and that she wanted her mother to make all medical decisions for her should she be unable to speak for herself. This note had an illegible signature AH's mother claimed was her daughter's signature. The attorney also stated that AH had withdrawn her advance directive when she was recently admitted to the hospital where she was now receiving treatment. AH's father was in attendance and spoke on behalf of his daughter's wishes, but he had not retained legal counsel and did a fairly ineffective job of describing what he thought should be done to keep his daughter comfortable while withdrawing life support.

The hospice representatives testified that AH was competent to make her own medical decisions when she was admitted to hospice and that she had executed the current advance directive upon admission to their care. They weakened their case, however, when they admitted that the original advance directive had been lost, and that while they believed the present document was similar, they could not testify that the two documents were identical. The judge who was presiding over the hearing announced that she "refused to be associated with another Terri Schiavo situation," and that life supportive care must be continued until the next hearing, which she then delayed for three months. She also named AH's mother as her guardian, despite what appeared to most observers as her erratic behavior, mental instability, and previous diagnosis of schizophrenia.

Meanwhile, the intensive care nurses and other staff members were forced to provide continued life support and care to AH that they believed she did not want. The consensus of the intensive care team was that continued life support for AH was futile, that her advance directive should be honored, and that life support should be withdrawn according to her stated preferences. Prior to the next scheduled court hearing, AH coded and died, despite the best efforts of the medical staff to revive her, having been maintained on a ventilator and feeding tube for three months.

Discussion Questions

1. The Terri Schiavo case presented a chilling precedent for end-of-life cases. Ten years after her death there were still those who felt her husband was right to advocate for what he believed (and convinced the court again and again) were her wishes not to be maintained on life support, and those who felt strongly that Terri was a disabled woman starved to death against her parent's wishes. Does the Schiavo case present a useful precedent for making ethical judgments in AH's case?
2. What are the pros and cons of seeking a court hearing in a case such as this? Are there alternatives that hospitals and physicians might pursue in order to ensure that patients' wishes can be followed without having to go to court?
3. "Slow code"[2] refers to situations where the "crash cart" team in a hospital or other medical center purposely responds in a slow or incomplete manner to a stat call to assist a patient in cardiac arrest; although rare, it may be done when CPR is judged to be of no medical benefit. Is this ethical?
4. Advance directives can be helpful under the right circumstances, and ineffective in others. How might advance directives be improved to avoid situations like this?

A Bioethicist Responds

Given the particulars presented in this case, it seems fairly astonishing that AH should end up in the situation she did at the moment of her death. First, the fact that this occurred in the same state as the Schiavo case means, according to Florida State Statutes, that both of AH's parents should have had equal standing as legal surrogates; furthermore, it occurred less than ten years later. Second, a significant irregularity appears to have taken place on the part of the attorney hired by AH's mother, one that goes unchallenged. Third, the judge appeared to have been more interested in avoiding a difficult family member and a potentially troubling public relations problem than in acting to do a right and good thing for an incapacitated patient. And, finally, the professionals caring for AH were destined to become frustrated when their combined skills, knowledge, experience and perspectives on situations such as those surrounding AH were effectively "high-jacked" by persons not having benefit of the same.

Admittedly, we do not know who AH might have named as surrogate, or surrogates, in her original advance directive, which the hospice reportedly had lost, so it is possible that her mother could have been named as her only surrogate. Nonetheless, the mother's determination to contradict AH's clearly stated oral preferences to her physician, which were corroborated in her advance directive, combined with her own history of schizophrenia, and her having hired an attorney who at one time had represented Terri Schiavo's parents, are problematic issues from the very beginning. Complicating the situation even further was this attorney's placement of a note in AH's medical chart claiming that she had the authority to revoke AH's advance directive. It is unclear where, in Florida Statutes at any rate, such a claim might be supported, and there is no indication in the case report as to why the judge would accede to this action. Moreover, it is irregular, to say the very least, for anyone outside those charged with direct, or delegated, care of a patient to add something to a patient's chart absent permission. The remaining description of the court hearing is equally troublesome, raising yet more puzzling questions about the judge's action, or rather, inaction in this case.

For whatever reason, the judge has chosen the easy way out here. It would appear that she had gleaned enough from the testimony of the medical team that AH would not survive an indefinite period of time, and so she delayed the case for three months, no doubt hoping that AH would succumb in the meantime. She obviously achieved her purpose while also sparing herself having to make any difficult decisions. It demands little imagination to suggest that she might also have thought she would be doing relatively little harm (not least to herself!) in making this choice, especially insofar as AH was in what was described as a persistent vegetative state. Yet she has, in fact, done a great deal of harm in that she has clearly not supported AH's right to have her advance directive followed, given no clear indications to the contrary that would, on the basis of valid evidence, justify an appeal to the standard of substituted judgment, or, failing that, the best interests standard. Moreover, significant harm is indeed done to the physicians, nurses and others who cared for AH by virtue of misuse of their time and skills, as well as offense to their professional integrity; to an already burdened health care system with limited resources and where continued life support had been judged to be futile by the intensive care team; and, most of all, to a patient who had indicated she did not want the treatment being administered, her lack of awareness notwithstanding.

AH's new physician, who came on the case at about the same time as the appearance of her mother's attorney, would have been entirely justified in discontinuing life support after the 30 days period had ended, and he was entirely right not to have been intimidated by the attorney's note in the chart, or the fact that she had worked on the earlier Schiavo case. It is very unfortunate, however, that the attending physician urged the hospital to seek the court's help in resolving what never should have been all that difficult a decision but which became complicated and was made potentially very troubling because of the weight of a notorious precedent case and the trauma suffered by many involved. Yet it is perhaps also unsurprising that he did so, or that he may have had no choice but to do so: hospitals generally have their risk management offices work closely with their ethics committees, often with an attorney representative of that office serving as *ex officio* member of the committee, and risk management could well have insisted the issue be referred to the court if for no other reason than to protect hospital interests.

What then happened, in effect, was that a person or entity (judge, court) other than the one with the true moral authority for the decision to discontinue treatment usurped that authority, chose poorly in making the decision (effectively no decision at all, as indicated above), did far more harm than good to all, and cast herself in a less than distinguished light in the process. While it is always easy to second guess after the fact, perhaps one approach the attending physician might have taken in attempting to reclaim some legitimate control over the situation, once it had become tied up in this judge's court, might have been to request an emergency review by a higher authority, a chief judge perhaps? Given the long shadow of the Terri Schiavo case, however, and particularly in the same geographic area where some of the same central characters were still living and working, nothing suggested by the legal system might have carried much weight.

The Terri Schiavo case[1] occupied the courts, newspapers, and ethics committees for 15 years in West Central Florida, and eventually became an international news story that involved the Supreme Court (they declined to hear the case), the Pope, the President of the United States, and the Governor of Florida. Her husband maintained, and was supported in every court hearing, that his wife would not want artificial nutrition and hydration to keep her alive in a persistent vegetative state, while her parents maintained that their daughter would never choose to starve herself to death, regardless of her condition. The judges and attorneys involved in the case were subject to death threats, harassment, and negative publicity. Terri Schiavo did not have an advance directive and had no knowledge of her medical condition; she lapsed into a coma and eventual vegetative state from which she never recovered and of which she was never aware.

AH's situation unfolded a few years after Terri Schiavo's death and in the same general geographical area, and no one wanted to be involved with another case in which another young woman would die because of legal action. So although AH's situation was markedly different in that she had managed a serious chronic illness for some period of time and had a valid advance directive, the evocation of Schiavo was enough to convince the judge involved in AH's case to prolong the proceedings until her death, rather than preside over what she felt might be "the next Schiavo case." Prolonging the next hearing is ethically questionable, but even more so was the appointment of AH's mentally unstable mother, who had made her intentions to thwart AH's advance directive abundantly clear.

The Schiavo case was a momentous one, and it became something of a political and religious issue attracting many impassioned voices. It is understandable how the judge in the AH case would thus be chary of presiding over "the next Schiavo." Prudence, therefore, also demands that a case such as Schiavo be used judiciously, along with other precedent cases, when attempting to resolve new ones. This is not to imply that Schiavo, or any other high profile case, should ever be viewed as so intimidating or as so exceptional that it be set entirely aside and not included with other precedent cases, regardless of how it may tend to reshape the paradigm, provided it is appropriate for inclusion with similar cases. It is to suggest, however, that those factors which might be seen to have attracted the stark attention to Schiavo, such as the political and religious debates that became partisan and rancorous at times, not to mention the death threats against the local judge who had ordered Terri Schiavo's feeding tube be withdrawn, be carefully "bracketed," or set apart insofar as possible, in order to promote an optimal, reflective and analytical response to the case at hand.

A Health Communication Scholar Responds

Law and ethics, as we mention in several of these case chapters, are different, but are hopefully able to work together to protect patients and to help ensure they get the medical treatment and legal protection that is most in line with their beliefs and preferences and state and federal statutes. The law failed in the Terri Schiavo case because it was not able to adequately address a situation in which a young adult patient's parents and spouse were so at odds. Even though Florida's end-of-life statute clearly identified Terri's husband (who was also appointed as her guardian) as her legal decision maker, it failed to provide guidance for situations in which a patient's parents disagreed and utilized the media and politics to press their point of view. The Florida statute assumes a fairly high degree of family harmony, which was certainly missing in both the Schiavo case and the case described here. The Schiavo case also occurred during a particularly conservative time in our nation's history, which also featured a U.S. President and Florida Governor who were brothers, and a Pope (John Paul II) who himself used a feeding tube at the end of his life.

All state laws governing end-of-life decisions must take the federal Patient Self-Determination Act (PSDA) of 1991 into account. The law requires hospitals, nursing homes, home health agencies, and health maintenance organizations (HMOs) that receive federal funds to routinely provide information to patients about advance directives at the time of admission. Advance directives allow a patient's autonomous choices about the medical care they desire to be maintained after they lose the ability to make or communicate their own health care decisions. Health care agencies must also develop policies and procedures to ensure patients' health care directives are followed; in Florida, health care providers who provide care in accordance with a patient's advance directive receive immunity from legal action. Most states, Florida included, suggest templates for advance directives that will be familiar to the health care facilities in their state or region, and it is to the patient's advantage to have an advance directive that appears thoughtful, official and complete. The scrap paper substitute that AH's mother produced in court should have been dismissed as fraudulent, but patients are not required to have advance directives that adhere to a specific format. A legally acceptable advance directive can be written (neatly) on paper, signed by the patient, and witnessed by a notary. A written document is more durable, but patients are also allowed to use oral statements as evidence of their advanced care planning and treatment preferences. In Florida, advance directives are defined as "a witnessed written document or oral statement in which instructions are given by a principal or in which the principal's desires are expressed concerning any aspect of the principal's health care."[3]

According to Florida law, a person's advance directive comes into play when they are unable to communicate their preferences for treatment, either because they are in a *terminal condition* (caused by injury, disease, or illness from which there is no reasonable medical probability of recovery and which, without treatment, can be expected to cause death); an *end-stage condition* (an irreversible condition caused by injury, disease, or illness which has resulted in a progressively severe and permanent deterioration, and for which treatment would be ineffective); or a *persistent vegetative state* (a permanent and irreversible condition of unconsciousness in which there is the absence of voluntary action or cognitive behavior or any kind, and an inability to communicate or interact purposefully with the environment) (Cranford 1988; Wade and Johnston 1999). Patients who have not been diagnosed with one of these conditions, or who are able to make their own decisions, are not held to the preferences outlined in their advance directives. For example, a patient who is expected to recover from his or her illness, disease or injury is likely to receive medical treatment regardless of instructions provided in their advance directives, and generally even if their surrogate or proxy decision maker demands that care be withdrawn. The preferred course of action in these cases is generally to treat the patient to the point where they are able to direct their own care. And patients who are competent to make and communicate their own treatment preferences can always override their written instructions even if they eventually meet the prognostic requirements for their advance directives to take effect.

A patient's advance directive, like AH's, is likely to indicate under what circumstances, if any, life-prolonging procedures are desired. Life-prolonging procedures specifically include not only ventilator support, but "any medical procedure, treatment, or intervention, including artificially provided sustenance and hydration, which supports, restores, or supplants a vital function."[3] Patients are free to choose the procedures and circumstances that conform to their values and preferences. The law states "every competent adult has the fundamental right of self-determination regarding decisions pertaining to his or her own health, including the right to choose or refuse medical treatment." It recognizes that life-prolonging procedures may result in "a precarious and burdensome existence," and that competent adults can make an advance directive instructing his or her physician to "provide, withhold, or withdraw life-prolonging procedures." As indicated in the case report, AH's advance directive did state her willingness to a 30-days trial of life supportive measures, after which they should be discontinued.

The Florida statute also outlines a hierarchy of proxy decision makers who can serve as a patient's decision maker, beginning with a legally appointed guardian, and then spouse, and ending with a close friend or licensed clinical social worker. The law further provides guidelines for reviewing the decisions made by a patient's proxy or surrogate decision maker (respectively determined by law, or by the patient's designation). These guidelines suggest that judicial review is appropriate if a proxy or surrogate makes decisions that are not in accord with a patient's known desires, has abused their powers or failed to discharge their duties, or if the patient's advance directive is ambiguous. It is not clear whether AH nominated her mother as her sole surrogate, or whether she intended her advance directive to dictate her care without much need for interpretation or advocacy.

There are problems with advance directives, and even well-crafted legislation can fail to address the practical realities of health care, or the twisted logic of human relationships. Advance directives can be lost, be incomplete or contradictory, be irrelevant to the situation at hand, or be contested by family members. It is impossible for a static document to communicate effectively all of the nuances of personal preferences for medical treatment, or to adapt to the exact conditions at hand. Laws can assume harmonious family relationships where they do not exist, and they can fail to account for the current reality in which everyone is entitled to their day in court. It is clear in this case that AH was in an end-stage and terminal condition and was perhaps soon also to be diagnosed as being in a persistent vegetative state as well (one need not be diagnosed as being in all three of these categories; just one will suffice). Her advance directive was completed upon her admission to a health care facility—a hospice agency—appropriately signed and witnessed, and then unfortunately lost. AH was also careful to specify the exact terms of her willingness to endure life-sustaining treatment. It would have been reasonable and perhaps even expected under the conditions of her hospice admission that she forego all life-supportive measures, and perhaps also to investigate whether having do-not-resuscitate status would further support her treatment preferences.

Many of us, however, may find ourselves in circumstances in which we make concessions to our loved ones, and AH did so by acknowledging her mother's disagreement with her refusal of life support. To allow her mother time to grieve, perhaps, or to say goodbye, AH agreed to no more than 30 days of ventilator and other life support; surely she did not intend for this concession to provide an opening for her mother to pursue legal action. As best as we can determine, AH had an advance directive that specified limits to her care, and she met the conditions under which it should be implemented. But similar to the Schiavo case, family disagreements circumvented the application of good law, sound ethical judgment, and expert medicine.

We might be tempted to say that AH's extended period of life support was necessary for her mother's grieving process, and that it did not significantly impact her due to her unconsciousness. Although many may feel AH's mother acted inappropriately, we must also keep in mind the particular difficulties of end-of-life decisions for chronically ill children who live until adulthood. AH was a chronically ill child whose mother had been making medical care decisions for her, at least for the first 18 years of her life. We are not privy to the process the two used to make medical decisions when AH was younger, but we do know their joint efforts resulted in keeping her alive to age 25. We also know nothing about the state of their relationship when AH suffered her final illness. AH's agreement to a trial of life support at least indicates concern for her mother's feelings and preferences. The process of assuming the right and responsibility to make her own decisions, about her medical care and other issues, was complicated by her long-standing illness. Her mother's serious mental illness also produced another difficulty in securing a smooth transition from parental decision-making to her own. Family dynamics are not accounted for in state laws governing end-of-life decisions, and the Florida statutes are no exception. Advance directives can be effective in specifying the treatment preferences of an incapacitated person, but they cannot completely substitute for a person's ability to speak for him- or herself or negotiate a compromise with a family member.

Despite the grief of family members, we are still entitled to have our wishes honored in spite of our incapacity, perhaps even more so. We also have to be mindful of the toll that providing futile or unwanted care exacts on nurses, physicians, and other health care providers, who experience moral distress and burnout when their job requirements are at odds with their moral judgments. What we can say about the hospice agency that "lost" AH's original advance directive and inadvertently supported her mother's off-base judgments is difficult to determine. We can only hope that the people we entrust to care for us, including our family members, our medical providers, and worst case, our legal system, will have our best interests at heart and will competently perform in their respective roles. All of these individuals, with the exception of the physician and the hospital who were willing to go to court on AH's behalf, failed her.

Notes

[1]Terri Schiavo was a young woman who was maintained with a feeding tube in a persistent vegetative state for 15 years while her husband and parents argued about her wishes for life-sustaining treatment. After trying aggressive and experimental treatment and rehabilitation for several years, her husband, Michael, maintained that Terri would not want to be kept alive in her condition with no hope of recovery; her parents argued that their daughter was a good Catholic who would never do anything intentionally to end her life. This case was the most litigated end-of-life case in history, and it ended in Terri's death in a hospice house after her feeding tube had been removed and reinserted three times. It was also one of the most highly publicized, emotionally charged and politicized cases in the history of American biomedical ethics. Much has been written about it, some of which lacks completeness and total objectivity. One starting point for a beginning look at Schiavo, and one which can be said to lay out the pertinent issues of the case as objectively as possible, without taking sides and with no hidden agenda, is Caplan, Arthur L., James. J. McCartney, and Dominic J. Sisti. 2006. *The Case of Terri Schiavo: Ethics at the End of Life*. New York: Prometheus Books.

[2]**Slow code** refers to the practice in a hospital or other medical center purposely to respond slowly or incompletely to a patient in cardiac arrest, particularly in situations where CPR is of no medical benefit.

[3]Florida State Statutes, Chap. 765. http://www.flsenate.gov/Laws/Statutes/2013/Chapter765/Part_I.

References

Cranford, Ronald. E. 1988. The persistent vegetative state: The medical reality (getting the facts straight). *The Hastings Center Report* 18: 27–32. https://doi.org/10.2307/3562014.

Wade, Derick T., and Claire Johnston. 1999. The permanent vegetative state: Practical guidance on diagnosis and management. *British Medical Journal* 319: 841–844.

Case 10—Please Stop Torturing Me! (Unless My Wife Is in the Room)

LA was a 55-year-old man who until recently had been in excellent health. He was active in many community business and philanthropic organizations. His first wife died of cancer several years earlier, which was traumatic for both LA and his adult son. LA remarried five years ago. Over the Christmas holidays he began experiencing drenching night sweats, unexplained weight loss, and shortness of breath and was diagnosed with Diffuse Large B-cell Lymphoma (DLBCL) shortly after New Year's Day. His oncologist was part of a community cancer practice with admitting privileges and patient care responsibilities at a large urban hospital. On January 19, one week after his diagnosis, LA presented to the Emergency Department, as directed by his oncologists, with severe abdominal pain. At that time he was hypotensive (low blood pressure), tachycardic (rapid heartbeat), and anemic (low red blood cells or hemoglobin) and had a large hemoperitoneum (blood in the peritoneal cavity) from a ruptured spleen, which is generally classified as a surgical emergency. LA was a very sick man.

LA was taken to Interventional Radiology for embolization of his spleen and was then admitted to the Intensive Care Unit (ICU) where he received large doses of blood products. He developed ischemic hepatitis, also called shock liver, which is characterized by sudden elevation—sometimes to as much as 20 times the upper limit of normal—of liver enzymes. In addition to systemic hypotension, a common accompaniment to shock liver, LA developed acute renal failure and elevated bilirubin, a possible sign of hepatitis or cirrhosis. His increasingly severe respiratory failure now meant he required ventilator support.

Over the next several days, LA was successfully weaned from the ventilator and was transferred to the Oncology Unit. He had his first course of chemotherapy on January 24. His wife and adult son were frequent visitors, and they closely monitored his treatment and well being. LA had pancytopenia (a reduction in the number of red and white blood cells), and he continued to be anemic and in need of frequent blood transfusions. LA received a second round of chemotherapy in early February. Shortly after, he was transferred back to the ICU so his altered mental status and recurrent respiratory failure could be addressed, and he was started on the

© Springer International Publishing AG 2017

L. A. Roscoe and D. P. Schenck, *Communication and Bioethics at the End of Life*,

https://doi.org/10.1007/978-3-319-70920-8_10

antibiotic Cefipime due to the presence of gram-negative rods. Gram-negative bacteria are resistant to multiple antibiotics, and are more typical in health care settings; they can cause infections such as pneumonia, bloodstream infections, wound or surgical site infections, and meningitis.

LA's altered mental status worsened, which may have been exacerbated by the administration of Cefepime in the context of renal failure. LA was then diagnosed as having myoclonic encephalopathy, an epilepsy syndrome that is resistant to treatment. He also developed a subdural hematoma with a midline shift, which is a collection of blood below the inner layer of the dura but external to the brain and the most common type of traumatic intracranial mass lesion. Mortality and morbidity rates are high, even with the best medical and neurosurgical care, especially when accompanied by midline shift. Midline shift is a shift of the brain past its center line, evident on neuroimaging such as CT scanning, and is considered ominous because it is commonly associated with a distortion of the brain stem that can cause serious dysfunction such as abnormal posturing and failure of the pupils to respond to light. Midline shift is also associated with high intracranial pressure, which can be deadly.

LA's wife and son were extremely concerned about his worsening condition. Since his initial admission was for DLBCL, the oncology group who had diagnosed his cancer oversaw his care in the hospital. His wife had become close to these oncologists, who were optimistic about their ability to cure LA's cancer, and who repeatedly reassured his wife and son that he would recover despite his increasingly serious complications. LA was also seen by the neurology and neurosurgery services, as well as by the intensive care team, while he was in the ICU. His wife was suspicious of these other physicians, however, and "fired" the first group of intensivists because she felt they were not hopeful enough about her husband's chances for recovery. They had encouraged her to talk with her husband about his treatment preferences, and even to consider a Do-Not-Resuscitate (DNR) order, since they had come to believe that continued aggressive treatment would be of unlikely benefit to LA. The information she received from the intensivists focused on LA's worsening, serious multiple organ failure rather than on the chances of his cancer going into remission, and this stood in conflict with the information provided by the oncology team. LA's wife refused to discuss the matter further and consequently refused to have the intensivists involved in her husband's care.

During this time LA's oncologists ordered a bone marrow biopsy, the results of which indicated that he was now free of lymphoma. His oncology team determined LA's cancer to be headed toward remission (which usually means being cancer-free for five years), his organ failure possibly reversible, and they recommended continued aggressive treatment for the other medical complications that had developed. All of this was good news to LA and his family, and as a result his wife became determined that no other conflicting information would spoil that view of things for her, her husband, or her step-son.

LA did not have an advance directive and at no time was he under a DNR order. In the presence of his wife and son, LA put on a brave face and reassured them that he was doing fine, that he was recovering slowly, and that he was greatly encouraged about his cancer's "remission." However, during the few times that his wife or son were not with him, LA told the nurses "I know I'm dying, please stop this torture!" When the nurses brought these comments to the attention of the oncology team, the oncologists interpreted this as evidence that, because of his neurological issues, LA was not competent to make his own medical care decisions. They relied solely on his wife for direction, who continued to demand aggressive treatment. The oncology team did not request either a psychiatric or neurological consult, which would have either confirmed or contradicted their opinion about LA's mental capacity.

LA's wife and son told the oncologists that they could not bear to lose him and that all available treatments must be tried. His wife provided consent to start LA on total parenteral nutrition, which dripped nutrition and hydration through a needle for 10–12 h each day. LA then developed a bowel obstruction, which caused extreme pain. His wife demanded that his pain medication be reduced when she was there so he could be awake and could interact with her; LA requested and received more aggressive pain medication when his wife was not visiting.

LA remained in the ICU, and even though the intensivists provided daily care, the oncologists, as admitting physicians of record, refused to collaborate with the ICU team about LA's care plan. The new intensive care team requested consults from the palliative care team, psychiatry, and the ethics committee. They were concerned that LA was not able to direct his own care, and that he may, in fact, be competent to do so. They also increasingly felt that continuing aggressive care was not in LA's best interest, especially since he had said on several occasions that his treatment was "torture." Nonetheless, the oncologists continued to direct his care, and they refused to allow any of the consultations to proceed. According to them, LA's cancer was in remission, and their patient and his wife had agreed to a plan of aggressive care to address his other medical issues. They documented that his bone marrow was clear of disease and that treatment should continue. The oncology team felt that some of the messages being conveyed to LA and his wife by the intensivists and ICU nursing staff were too negative, unhelpful, and not professional.

The chaplain from the palliative care team was able to visit with LA during the few times when his wife was not present, and talked with him about what he perceived as his impending death. The chaplain's notes of March 9 read: "Pt is tired of fighting and says he wants to pass peacefully and no longer wants to go on living if this means going through all these medical procedures." The chaplain tried on many occasions to meet with LA's family, but they rejected pastoral support and refused to meet with anyone but the oncologists. On March 16, the chaplain again met with LA who said, "I wish I could close my eyes and this would all be over." The next day LA and his wife were together, and on March 17 LA stated that he had hopes for a full recovery and would never make any decisions that would run counter to his wife's or son's wishes. "I just want to sit in a chair and hold my wife," LA said, "and I will do anything if it makes this easier for her." Over the course of that month, LA was transferred back and forth between the ICU and the oncology unit. He was once again transferred back to the ICU on the evening of April 3 to treat rapid atrial fibrillation. A few hours later the nursing staff found him unresponsive and called a full code, which involved chest compressions, electric shocks, and the administration of emergency medications. Despite these efforts, LA did not survive and was declared dead on April 4 at 12:30 am.

Discussion Questions

1. How can hospitals ensure that consistent messages are conveyed to patients and families when several medical services are involved in a patient's care?
2. How should medical professionals react when ethics and other consults are rejected by family members or by other health care providers?
3. How might this patient and family have been counseled to understand that while the primary cancer diagnosis was responding to treatment, the patient's condition remained extremely poor?

A Bioethicist Responds

Those who care for patients in ICUs, those who care for the terminally ill, and even family members or friends of those who have died after long illnesses in hospitals or long-term care facilities may not find this case all that extraordinary. The particular circumstances may well seem egregious, yet the general situation presented in the foregoing narrative might also remind one of similar cases wherein patient capacity appears never to have been clearly determined, if at all, and where in the absence of such determination family members and/or attending physicians have decided that patients lacked capacity to make their own medical decisions. LA's story is certainly not unfamiliar to the present writer, who has served on a number of hospital ethics committees over a period of twenty-five years.

The narrative is clear that LA had no advance directive. It may also be presumed that at some point during his hospital admission, and allowing only for the possible failure of Admissions Office personnel to observe the requirements of federal law, either he or his proxy would have been asked if he had an advance directive or would like to prepare one at that time (Volandes 2015). But, we do not know for sure what may or may not have transpired with regard to this issue at time of admission. Still, it is curious that LA apparently never asked that anything be put in writing regarding his wishes concerning medical treatment, despite his apparent clarity of mind when speaking with the nurses and hospital chaplain. Perhaps the fact that LA never committed anything to writing, never prepared a Living Will or Advance Directive, never signed a Durable Power of Attorney or designated a health care surrogate, speaks to his age and relatively good health heretofore. He is undoubtedly like many, if not most, persons his age who have yet to encounter serious illness: "young," at least from a medical point of view, which is to say, still in the prime of his life, well and, therefore, unlikely to be particularly concerned about end-of-life issues; he was still active in business and the community; and, having remarried only five years prior, he must surely have been thinking of a second life plan. The health crisis that apparently beset him quite suddenly, with little or no warning, could have only had the effect of turning his world completely on its head. Few persons in this situation are likely then to pause, take stock of who and where they are, and to begin the careful, reflective process of planning for their future with family, physician(s) and friends, a process that ideally would include discussion of advance health care planning, living wills, durable powers of attorney, surrogacy, as well as the drafting and filing of appropriate documents. Moreover, it is certainly understandable that a sick, very vulnerable patient would entrust his care and well being to his closest family member and physician at this critical time. At least in the beginning.

The dynamics of this case are complex, though, again, not all that unusual. LA's wife seemed genuinely interested in her husband's welfare, determined as she was to share his oncologists' optimism that he could be cured, even to the point of "firing" the first group of intensivists who encouraged a different treatment approach, yet she quickly began driving the action and directing treatment decisions. The oncologists in turn seemed only too willing to take their cues from her to the point of refusing to collaborate with the intensive care team and refusing to allow any requested consults to go forward. Meanwhile, LA acquiesced to the decisions made for him by his wife, all the while complaining privately to his nurses and the chaplain about how he was being treated. It is unlikely that LA was simply "giving in" to his wife or that he really was so incapacitated that he was incapable of speaking rationally and competently for himself. The testimony of the chaplain should be sufficient evidence to the latter. Furthermore, LA's apparent acquiescence before his wife is most likely explained as the terminal patient's desire to be the strong one, the one to hold out hope for those around him who are undergoing their own suffering and who must struggle with their own anxieties and fears about death and loss. LA made it clear that he knew he was dying and that he wanted his suffering to be over, yet at the same time he had stated unequivocally that he would never have made any decisions counter to his wife's or his son's wishes and had stated, "I will do anything if it makes this easier for her." Were this the complete picture of LA and the dynamic in which he is found at the end of his life, he might almost be seen as a rather noble figure of a man.

Yet this is not the complete picture, which is made infinitely more complex by the behavior of the oncologists, such that in the final analysis LA appears largely as victim, the unfortunate subject unjustly deprived of his rights, autonomy and freedom so that the world according to his wife and those physicians earning her approval may proceed with their own narrative. This is more than a simple deprivation of rights, autonomy and freedom, however; the violation of at least one state law, or statute, appears to occur as well.

The oncology team relied solely on LA's wife for guidance in terms of treatment decisions and chose unilaterally not to seek confirmative consults for their determination that LA had lost medical decision making capacity. While attending physicians are not _required_ by Florida state law to obtain confirmatory consults in situations regarding capacity, seeking one in this case would certainly have been the most prudent approach. LA's oncology team chose to interpret the nurses' report of LA's wish to die as evidence of medical incapacity. If there is any truth to this, it is evidence only of strong paternalism requiring justification, and begs for confirmation upon the general assumption that not everyone would possibly draw the same conclusion to a patient's expressed wish for his suffering to end. Florida's end-of-life statute, F.S.765.204, states: "If the evaluating physician has a question as to whether the principal lacks capacity, another physician shall also evaluate the principal's capacity, …"[1] It might be argued that LA's physicians had no question whatsoever concerning his capacity, but whether this argument could be made in total candor seems open to doubt given the exclusive relationship that developed between them and the patient's wife. The fact that the attending physicians chose not to ask for confirmation of their assessment of the patient's capacity demonstrates at least a violation of the spirit of this statute.

Regrettably, LA's wife also violated one of the fundamental principles of biomedical ethics in demanding that his pain medication be reduced when she visited him in the hospital, the principle of nonmaleficence.[2] Her behavior in this regard, while understandable from a certain perspective given the stresses upon her, is primarily egocentric and self-serving.

LA suffered needlessly. This is not to say that anyone wished him harm or wanted him to experience pain or discomfort. His family had difficulty seeing themselves and their own suffering over LA's impending death as different from his, while at the same time the oncology team seemed to have difficulty relating to LA in caring, patient-centered ways. It would have been unlikely that a resolution to this situation could have developed out of the patient-physician-family triad. Where it might have been found, however, is from within that other group of persons privileged to have a close relationship with LA, that is the nurses and the chaplain. These persons can typically be real patient advocates. The only problem is that all too often they are not empowered to speak up or to be the advocates they have a legitimate right to be for the patients they serve. A team effort is required to serve patients properly, especially in complex hospital settings today, and all members of the team should feel comfortable playing their roles in responsible ways. All do not have equal responsibilities here, but none should feel so disenfranchised that they must remain silent when genuine concerns of real value to patient care and well being are at stake. And above all, it must be remembered that no one, nor any group of persons, should ever have the power simply to veto or forestall legitimate processes, such as determinations of patient decision-making capacity and ethics committee consults, designed to promote beneficence.

A Health Communication Scholar Responds

It can be bewildering for patients and their families when a formerly healthy person is diagnosed with a serious illness, especially when the outcomes of treatment are ambiguous. In LA's case, he was in the unenviable position of being cancer-free, but in multi-system organ failure. The extent to which his cancer treatment caused or contributed to the complications he experienced is an open question. Cancer treatment by its very nature is destructive; our hope is that we are increasingly able to target its destructive power to cancer cells alone, sparing healthy tissue and function.

It is quite easy to become frustrated with LA's wife and her insistence on continued aggressive treatment and her inability to hear information contrary to her wishes. It is natural for patients and their families to cling to good news and to those who convey it, and to shun those who provide information that complicates or nullifies the "good news" they think they have received. One must try to be sympathetic to the plight of LA's wife as she tries to come to terms with her husband's increasingly dire medical condition. It is possible that she did not understand the ways in which various medical services attempted to engage her in discussions about her husband's care, and believable that she would prefer to let the oncologists and their hopeful message of recovery guide her decisions. This case reads like an exaggerated version of a familiar, albeit misleading scenario, in which a health care provider comments on a subtle improvement in a patient's blood chemistry, for example, while failing to emphasize that the patient's overall condition and prognosis remain dire. No one likes to be the bearer of bad news and no health care provider wants to believe the treatments they ordered for their possible benefits might ultimately prove futile or even harmful.

When a patient has a serious and complicated diagnosis, the ways in which information is provided to the patient and family becomes an even more crucial element in formulating a treatment plan. Good communication practices can also help facilitate coordination between medical services, and can help patients voice their own treatment preferences even if they do not align perfectly with family members' hopes or expectations. We have all likely heard stories that highlight the lack of coordination between medical services, and about the fragmentation that continues to characterize the American health care system. The lack of communication and coordination between the intensive care service and the oncologists involved in LA's case reached epic levels. Physicians of various specialties are generally courteous to one another and generally hesitant to call one another out for bad behavior, but it appears clear from the case description that the oncology team overstepped their boundaries in excluding the expertise of the intensivists, the psychiatrists, the neurologists, the palliative care team, and the ethics committee. It is unrealistic to think that LA's wife would be capable or willing to attempt to untangle the roots of what must have been a long standing series of conflicts between the medical services involved in her husband's care.

At times LA and his wife appear to be in agreement that aggressive treatment should continue, and at other times, LA made it clear to the health care team that he knew he was dying and hoped they would stop "torturing" him with continued treatment. Had LA's wife allowed the palliative care team to become involved, perhaps they would have been able to facilitate a conversation between husband and wife to resolve their contradictory preferences and expectations. While we tend to regard autonomy as pertaining to an individual, some accommodations need to be made in order to honor and respect an individual patient's treatment preferences and also acknowledge that the patient wants also to honor his or her spouse's or family members' preferences too. Palliative care team members are experts in helping patients and families negotiate and decide on a course of action, even when all the available choices appear to be bad. Unfortunately, "palliative" is associated with dying; maybe another name could be given to the palliative care service to make it easier for patients and families to take advantage of their expertise. Euphemisms do not always work, of course, and it is likely in this case that LA's wife would not have allowed anyone in her husband's hospital room who thought talking about death or a change in the course of treatment was necessary, or even more effective pain management.

Recent changes to Florida's statute directing end-of-life care in some ways further confound the issues in LA's complicated case. A new provision in the law allows a person to choose to have their designated health care surrogate make all medical decisions for them even if they retain capacity.[3] Given LA's acquiescence to his wife's demands for aggressive treatment, it seems likely that he would have chosen the option of allowing her to make medical decisions for him even if he was competent to do so himself. The language in the statute still privileges any decisions made explicitly by the patient, even if the surrogate disagrees, but provides little guidance in a situation such as this where the patient refuses to express treatment preferences unambigously.

This case took a tremendous toll on the ICU team. The ICU team members—the second team after the first group was "fired" by LA's wife—were very supportive of the family but predictably did not react well when the family began dictating care, which was particularly problematic since the health care team members were aware of the discrepancies between LA and his wife. LA's wife was constantly critical of the ICU team, and at one point demanded that his blood be redrawn three times because she "disagreed" with the results and felt the nurses must have done something wrong. The ICU team members were concerned that the oncologists had not helped the patient or family understand the gravity of LA's medical condition overall. The cancer diagnosis may be responding to treatment, but the rest of LA's body was rapidly failing. The ICU staff truly felt they were torturing their patient, and several nurses were in tears over what they were required to do to LA to comply with medical orders.

At some point, LA's cancer diagnosis became the least of his problems, but organizational procedures kept the oncology team as his treating physicians, and the cancer diagnosis in the forefront of his wife's consciousness and decision making. It is completely understandable that patients and families prefer good news, but the extent to which LA's wife and the oncology team were able to silence competing opinions about his condition is unconscionable. All medical professionals want to claim victory, but in LA's case, the "win" over his cancer was far overshadowed by the extreme and ultimately fatal complications he endured.

The reality of current day hospital administration is that specialist medical services are often contract employees, and the institution responsible for patient care is sometimes held hostage to the demands and dictates of physician practices who are free to take their expertise and patient caseloads elsewhere. The communication challenge is obvious: Consistent messages should be conveyed to the patient and family to allow decision making to proceed in accord with new information, but there are not always organizational mandates that require all those involved to meet and agree on the patient's status, recommended treatment plan, or the language chosen to describe both.

Another serious difficulty here is the extent to which the oncology group directing LA's care was able to shield their patient and his family from competing concerns about his care and treatment goals. In theory, anyone can call an ethics consult, but this freedom is severely curtailed if the patient's treating physicians can forbid new information from being considered. LA's own voice was also silenced. How are we to manage patients who tell us one thing in private, and then publicly acquiesce to the treatment preferences of their family members? LA's wife essentially trampled over his preferences for treatment and the information from the intensivists as well as the other relevant services—chaplaincy, palliative care, ethics, psychiatry—that had contributions to make that would have ameliorated LA's distress, as well as that of the nursing staff in the ICU (and probably of LA's wife and son). We cannot fault LA's wife for wanting to believe only hopeful messages, but we can fault the organizational structures that allowed her to only hear what she wanted to hear.

The moral distress experienced by the nurses and intensivists in the ICU is a real issue that needs to be addressed (Corley et al. 2005; Elpern et al. 2005; Hamric and Blackhall 2007). State law pertaining to this case allows physicians to withhold care they believe to be futile, but since futility is always a value judgment, it is extremely rare for a physician to withhold treatment on this basis (Schneiderman et al. 1990). Treatments that were keeping LA alive at great burden to him and enormous distress to some members of the medical staff were not futile in the opinion of his wife (or of the oncologists). Medical staff members, particularly nurses, bear a high emotional toll when compelled to do procedures they feel are more burdensome and harmful than beneficial. And having to navigate a patient care situation where the patient says one thing alone and another in the presence of his wife is especially taxing.

The other long-lasting outcome of LA's case is the continuing animosity between the intensivists and the oncologists at this large urban hospital. It is an understatement to say that the inability of these physicians to meet, talk, and agree on a course of action is disappointing. The ICU team felt that they were thrown under the bus by the oncologists, who wanted to celebrate their short-lived victory over LA's cancer. The absolute power of the admitting physicians to dictate a patient's course of treatment merits reexamination, especially since it is very likely that these physicians will be involved together in other complicated patient care situations in the future.

Notes

[1]Florida State Statues, Chapter 765.
http://www.flsenate.gov/Laws/Statues/2013/Chapter765/Part_I

[2]In the authors' opinions, the best treatment of the well-known "principles" of biomedical ethics is to be found in: Beauchamp, T.L. & Childress, J.F. (2013) *Principles of Biomedical Ethics*, 7th ed. New York: Oxford University Press.

[3]The new language states: "My health care surrogate's authority becomes effective when my primary physician determines that I am unable to make my own health care decisions **UNLESS** I check and initial one or both of the following boxes:

If I initial this box, [ ] my health care surrogate's authority to receive my health information takes effect immediately.

If I initial this box, [ ] my health care surrogate's authority to make health care decisions for me takes effect immediately, pursuant to section 765.204(3), Florida statutes, any instructions or health care decisions I make, either verbally or in writing, while I possess capacity shall supersede any instructions or health care decisions made by my surrogate that are in material conflict with those made by me."

References

Corley, Mary C., Ptlene Minick, R.K. Elswick, and Mary Jacobs. 2005. Nurse moral distress and ethical work environment. *Nursing Ethics* 12: 381–390.

Elpern, Ellen H., Barbara Covert, and Ruth Kleinpell. 2005. Moral distress of staff nurses in a medical intensive care unit. *American Journal of Critical Care* 14: 523–530.

Hamric, Ann B., and Leslie J. Blackhall. 2007. Nurse-physician perspectives on the care of dying patients in intensive care units: Collaboration, moral distress, and ethical climate. *Critical Care Medicine* 35: 422–429.

Schneiderman, Lawrence J., Nancy S. Jecker, and Albert R. Jonsen. 1990. Medical futility: Its meaning and ethical implications. *Annals of Internal Medicine* 112: 949–954. https://doi.org/10.7326/0003-4819-112-12-949.

Volandes, Angelo E. 2015. *The conversation: A revolutionary plan for end-of-life care*. New York: Bloomsbury.

Case 11—Who Should Make Treatment Decisions for a Battered Spouse?

LF and her husband WP were born and raised in Guangdong province in China and relocated to Florida soon after their marriage. This was a second marriage for both. LF had never been accepted by her first husband's family, especially once she began complaining about her husband's bad temper and occasional violence. Continuing disagreements between LF and her in-laws were the cause of the divorce; it is not unusual for a Chinese husband to side with his parents in the case of quarrels between them and his wife, even to the point of divorce.

LF had a 12-year-old son from her previous marriage who remained in China with her extended family when she and WP moved to the United States. They settled in West Central Florida, rented a small apartment, and joined a distant cousin of WP's who owned a Chinese restaurant. Her husband soon took over as manager, and LF, now pregnant with her second child, worked from home to balance the restaurant's books and manage the payroll. LF turned all of her earnings over to her husband, and except for brief trips to the restaurant to pick up paperwork, spent her days in their apartment. LF's husband was not pleased with the news about her pregnancy, as he had hoped to achieve greater financial stability before they had a child together. He began calling LF offensive names, and closely monitored her conversations with her family in China and the few restaurant employees with whom she had become friendly. Two of the waitresses wanted to hold a small party to celebrate LF's pregnancy at the restaurant before they opened for dinner, but WP would not allow it. "Your 'friends' think we're too poor to be having another baby!" he yelled. "They are right but you have no business telling anyone what goes on in our family!" It was more likely that the waitresses only wanted to acknowledge what they assumed was the happy news about the new baby. LF was disappointed about not being able to have a small celebration, which would have been a welcome break from working on the restaurant's books and her responsibilities at home. Baby showers are not common among Chinese families, many of whom are superstitious about celebrating the baby before he or she is born. "It's just as well," LF thought to herself, "no need to create problems with my husband or tempt fate".

© Springer International Publishing AG 2017
L. A. Roscoe and D. P. Schenck, *Communication and Bioethics at the End of Life*,
https://doi.org/10.1007/978-3-319-70920-8_11

LF had little direct contact with her older son, and relied on infrequent telephone calls with her parents and other relatives in China who reported on his well being. Although she missed him, she believed it was in his best interest to remain in China to complete his secondary schooling, and she hoped he would choose to come to America when it was time to enroll in college. Her son had no contact with his biological father, who had also remarried and fathered another son. Although her second marriage was more stable, LF had unknowingly married another abusive man. LF did not perceive either her husband's verbal name-calling or controlling behavior as abuse, and accepted it as something she had to tolerate. She had already brought considerable shame to her parents in China when she divorced the first time, and she was not willing to risk the social and familial consequences of another failed relationship.

LF gave birth to her second son and continued to work from home to help with the restaurant. The economic downturn of 2008 drastically affected the profitability of the restaurant, which was located near a new residential community still under construction. When the housing bubble burst, new housing starts declined precipitously, along with the restaurant's business. Pressures mounted on WP to maintain a stable income for the family, especially with the addition of the new baby. He took his frustrations out on LF, and the verbal abuse that had begun when she was pregnant escalated into physical violence. Although they needed the additional income, LF was often unable to go to the restaurant to pick up paperwork or to make bank deposits, because her bruises and abrasions were increasingly difficult to explain.

LF had by this time lost almost all contact with her family in China. She tolerated her husband's abuse, possibly to prevent him from taking out his bad temper and frustrations on the new baby. In spite of the violence she increasingly experienced, LF was determined to make this second marriage a success. As the economy slowly recovered, business at the restaurant gradually improved, but her husband's abusive behavior continued to escalate. He alternated periods of kindness and remorse with increasingly violent attacks against LF. By the time their young son was 5 years old, WP was threatening LF with hits, kicks, shoves, slaps and chokes, often in full view of their son. LF knew that although her husband was not abusing their son, witnessing the violence she endured was also harmful to him. Her older son was now ready to attend college, and she encouraged him to apply to the university in the city where they lived; it had recently expanded its outreach to international students, and this included additional scholarship opportunities and extensive English language training.

LF's older son was accepted into the university program and he arrived in his new hometown the summer before classes were to begin. His English was rudimentary, and his knowledge of American customs, not to mention the norms of American family life, was almost non-existent. LF hoped that her son's arrival would signal the end of the abuse she had endured, since there would be another nearly grown man in the home to witness the violence and to protect her. Her husband was unhappy his step-son would be spending the hot summer months in their cramped apartment, as he had expected him to move directly into his dorm room in time for fall semester classes. WP was brusque with his step-son and only spoke to him in English when he bothered to speak to him at all.

Shortly after the older son's arrival, LF's husband shoved her into a wall in the kitchen because he was unhappy that she had brought food home from the restaurant for dinner rather than cooking something from scratch. She fell against some drinking glasses sitting on the counter, and several of them crashed to the ground. The older boy was stunned, and his step-father screamed at him to stay away. The younger boy, a terrified 5 year-old, stepped between his mother and father, who grabbed a piece of broken glass and sliced his young son's finger. He then barked at the older boy to clean up the mess and fix his brother's finger. He led his wife into their bedroom, locked the door, and proceeded to beat her to unconsciousness.

Both boys were distraught, but were largely unable to communicate since the younger boy spoke only English and the older boy mainly Chinese. Neither of the boys were prepared to call for help—they were not familiar with "911" and the family had no close friends or neighbors. In any case, the idea of airing the family's dirty laundry was unthinkable. After 45 min of terror, WP carried his unconscious wife to their car, and shouted at the boys to remain in the house and instructed them to talk to no one.

Upon arriving at the Emergency Department of the nearest hospital, WP told a story about coming home from the restaurant and finding his wife unconscious and his younger son injured, and he made some vague references to the fact his stepson was in the apartment at the time and appeared uninjured. The medical team informed WP that his wife's condition was extremely serious; she had severe bleeding from the brain and a broken wrist. Within a few hours LF became comatose and never regained consciousness. Despite the fact that the lapses in WP's story rendered him a less than reliable narrator, he had no criminal record since LF had never reported his abuse to the police. Although LF's injuries were consistent with abuse, the hospital personnel had no immediate choice but to rely on WP to make medical decisions for his wife. As her condition continued to deteriorate over the next few weeks, WP insisted on continuing life support, and LF was kept on a ventilator and feeding tube even though the medical team all agreed her prognosis was extremely poor.

Meanwhile, the older son was able to communicate with his new University about the troubled state of his family life. With the administration's help, he filed a police report detailing the events of that tumultuous night. As a result, the younger boy was removed from the home and placed in foster care, and the older boy was allowed to move into his dorm room earlier than planned. The physicians in charge of LF's care knew that continuing to rely on her abusive husband's decisions about her medical care was a grievous error, but they were told by the court that until the sheriff's office arrested WP, there was nothing they could do to remove him as her legal proxy. WP did not want his wife to die, since then he would likely be charged with murder in addition to child endangerment and domestic abuse. The doctors now knew of LF's older son's existence, and attempted to contact him and convince him that he was the better proxy decision maker.

LF's older son, although extremely distressed about his troubled introduction to American life, his mother's poor condition, and his step-father's violence, had no knowledge of what his mother would have wanted in terms of medical care. He pleaded through a translator that he did not want to make any decision that would look as though he was the cause of his mother's death, and also that he did not know her well enough to know what she would have wanted. LF remained on life support, and the hospital petitioned the court to appoint a guardian. Before one could be appointed, LF succumbed to her life-threatening injuries and was pronounced brain dead after 6 weeks in the Intensive Care Unit. WP was arrested and eventually charged with murder, child endangerment, and domestic violence, and he was ultimately sentenced to life in prison. LF's older son withdrew from the University and returned to China, and he was distraught that he could not bring his younger half-brother with him, who remained in foster care.

Discussion Questions

1. Law and ethics should work together, but are sometimes at odds. What are some of the ways in which what was legal and what was ethical were problematic in this case?
2. Who should make medical decisions for a family member when criminal behavior is suspected but before criminal charges are filed?
3. Can other family members who appear to be more suitable decision makers for incapacitated family members be pressed into service in spite of language and cultural differences?

A Bioethicist Responds

The preceding narrative bears a major characteristic often found in case reviews, that is, it lacks information that might obviate questions such as: What was the dynamic between the hospital, the court, and the sheriff's office that permitted WP to act as his wife's proxy, despite suspicions that he was the cause of her injuries? And what resources, if any, were brought in to help LF's older son understand his decision making options, or to help stabilize the living situation of both of LF's sons during her hospitalization? Although one could always wish for more information than might be given in a particular case report, much can still be said about how the frustrating issues in this situation might have been approached differently.

LF and her children faced many obvious problems, such as spousal abuse; psychological trauma; fear (of reprisals, of authorities in a strange land, etc.); and familial embarrassment, shame, and dishonor. LF, the other members of her family, and everyone attempting to help her medically or to resolve the attendant social problems were operating in a zone to which two cultures—Chinese culture and the culture of American medicine—both laid claim. Culture is a set of shared patterns of behaviors, interactions, and cognitive constructs learned by members of a social group through exposure and socialization. Cultures and subcultures have defining ideals and norms that are unique. For example, our American medical/ethical focus on autonomy can be seen as a logical extension of our reverence for individualism, self-determination, independence, and freedom. Broadly speaking, Asian cultures favor a communitarian way of being in the world, and view the family as more important than any individual's desires or aspirations; filial piety is prized above the ability to make one's own way.

There are many instances of the culture clash between patients and the medical professionals attempting to provide care that have been documented. One excellent example is Anne Fadiman's book that explored the clash between doctors at a small county hospital in California and a Hmong refugee family from Laos over the care of Lia Lee, a child diagnosed with severe epilepsy (1997). Lia's parents and her doctors both wanted what was best for Lia, but the lack of understanding between them led to tragedy. Lia's illiterate parents were not able to give her the medicine prescribed for her condition, and could not accept the idea that she would need to take medication for the rest of her life. They also saw her seizures as evidence of her special connection to the sacred. The case of Lia Lee and the case presented here of LF—and many others—ask us to consider whether there are human ideals that transcend culture. Must we allow an epileptic child to be grievously injured because her parents' cultural beliefs see epilepsy as a divine blessing? Can we stand by and ignore an immigrant woman's suffering and abuse because to witness it or to intervene brings shame to her family?

The ethical issue with LF centers on the potential disconnect between Western medical views of autonomy and the Chinese cultural value of family and community. In the United States, we attempt to act ethically by honoring an individual's autonomy, operationalized in a medical context as his or her preferences for medical treatment. In LF's case, her cultural values placed the needs of her family ahead of what might best support her individually. Her willingness to stay with her abusive spouse speaks to this cultural value, as well as to more practical issues such as how she perceived her options for seeking assistance. Does honoring LF's autonomy and preference for self-sacrifice mean her abuse should be allowed to continue? We must ask whether it makes sense to hold fast to this relatively narrow view of autonomy, especially when the person's life has been endangered by someone usually identified as the proxy decision maker most able and willing to support the incapacitated person's notion of autonomy.

The principle of autonomy is a subset of a more inclusive ethical principle usually called "respect for persons" (Veatch 2003). Respect for persons is an ethic that derives in large part from the work of the philosopher Immanuel Kant. Kant affirmed the intrinsic value of human life and that humans deserve respect, demonstrated in the duties we have toward one another. The duty-based principle of "respect for persons" usually includes autonomy, fidelity, veracity and avoidance of killing. "Avoidance of killing" can be interpreted as a medical act—not participating in euthanasia for example—as well as literally not killing one another. Surely one of our most basic duties toward one another is the assurance of one another's personal safety and bodily integrity.[1] Our most basic ethical commitment must be to safeguard LF's life, even if such actions run counter to the patient's and family's cultural values. Culture is tricky. We need to explore the damage we do when we insist on the superior humanity of our own preferred cultural practices and ethical principles. And we must ask ourselves what harm we inflict when we look the other way from cultural beliefs or practices that injure, demean and sometimes kill.

While this case did not occur in an area with a sizeable Chinese population, it did occur in a very large, culturally diverse, highly educated, metropolitan area containing a number of colleges and universities, including a major state research institution. The area in which it is located is one where speakers of major dialects of Chinese could undoubtedly be found, where certified interpreters would no doubt also be available, and where there would surely be persons capable of serving as informants about Chinese culture and customs for Americans working in health care, social work, legal and law enforcement activities. At this point it may be unfair to judge those persons directly involved in this case (e.g., physicians, nurses, social workers, police), but in the absence of other information one cannot help wondering why there was no attempt to better understand LF's domestic situation, and the cultural values that kept it invisible before her untimely and tragic death.

A Health Communication Scholar Responds

The individuals in this case cannot literally communicate with one another, either because they do not speak the same language, or because their cultural values are so disparate as to be nearly unbridgeable. The husband and sons will not talk to hospital personnel or to law enforcement. LF can no longer speak for herself because of her severe injuries. Law enforcement personnel are not communicating effectively with hospital administrators. The two half-brothers barely know one another and do not share either a common native language or culture. WP and LF did not communicate effectively in their lives together, and WP used his cultural privilege as head of the household to abuse his wife and step-son, without any legal repercussions (at least initially). The legal right of a spouse to act as his or her partner's proxy decision maker is based on the assumption that they will act to support one another's best interests, which also assumes adequate communication between husband and wife. The patient's older son has some rudimentary English language skills, but is a world away from embracing his role as an adult son who can (and probably should) question his step-father's actions and decisions. The entire family is thrust into a specialized subset of American culture—that of Western medicine—which further stresses their ability to communicate. Even more complicated, perhaps, are the multiple roles that each must enact. Chinese culture might describe the role of the dutiful step-son, the powerful patriarch, the humble wife and mother, but American culture, law, ethics, and medicine require a decisive step-son, an apologetic husband, and an autonomous patient. Under the most stressful circumstances imaginable, this family is attempting to manage multiple identities, play multiple roles, and satisfy multiple audiences. And the health care team members caring for this dying woman are also attempting to understand the culture that this family carries with them, respect the law that governs medical decision making for incapacitated patients, and provide the best medical care possible. It is not clear from the case description how clearly and consistently the medical team documented the futility of continued life support for LF and pushed for a resolution to her medical condition, or how actively they advocated for more information about the cause of her injuries, so the areas in which communication gaps occur are multiplied.

In addition to the interpersonal and institutional communication difficulties these individuals faced, there are communication difficulties at a systems level between this immigrant family and the community in which they lived, worked, and raised their family. The communication discipline is particularly attuned to issues of context at multiple levels, and trying to understand some of the cultural differences Chinese immigrants face in adjusting to American culture might shed some light on this unfortunate set of circumstances.

The Pew Research Center reports that Asian Americans are the fastest growing racial group in the United States, comprising 36% of all immigrants who arrived in 2010; Latinos were second at 31% (2013). At 23%, Chinese Americans constitute the largest portion of the Asian American population. Nearly 2.4 million ethnic Chinese are living in the United States, and 2.2 million speak Chinese as a first language. Many reportedly have no intention of remaining in the U.S. permanently, and they work hard to maintain cultural ties and Chinese language fluency. Historically the primary immigrants from China have been men, with female immigrants entering the U.S. legally dependent on men as their wives, daughters, mothers, or sisters (Dasgupta 2000). Legal dependency often translates into financial and emotional dependency in South Asian families. Chinese men generally control the family finances regardless of a woman's employment status, and women who work outside the home are often perceived to be doing so as part of their domestic duties rather than for personal or professional reasons.

The image of South Asian femininity includes submissiveness and putting one's husband and family ahead of one's own needs. A woman may be educated, but that accomplishment brings status for her parents and future in-laws rather than for herself or her financial or emotional independence. Sons are valued more highly than daughters because they carry the family name, as well as whatever money, power, and status are handed from father to son. The importance of lineage is a distinct feature of Chinese culture. Daughters are seen as a financial burden to their families since they are expected to marry and become members of their husband's families. Chinese girls are taught from an early age that great harm will be inflicted on the family should they fail to be good wives. Even so, the marital relationship between a husband and wife is seen as secondary to the relationship between a son and his parents.

While often regarded as a model minority, domestic violence is a pervasive problem among Asians residing in the United States (Midlarsky et al. 2006). It is difficult to report accurate assessments of the prevalence of domestic violence in the Chinese immigrant community. The data available reveal between 10 and 25% of Chinese Americans are physically abused by a partner. But such behaviors as having one's earnings taken, being confined to one's home when not at work, or being belittled by one's husband are not always considered forms of abuse or violence. Among Chinese people, one is rarely seen as independent or separate from one's family or community, and this is especially true for women. Domestic abuse is viewed as a family issue, and the priority is to spare the family public humiliation or legal repercussions as opposed to protecting the victim (Ahmed et al. 2001). Chinese Americans largely agree that making domestic violence public violates family privacy and brings shame to the family. In addition to informational and language barriers, these cultural values impose additional obstacles to immigrant Chinese women who seek protection from violent partners.

LF was fluent in English, but she was also likely prevented from seeking help in part by these cultural norms. The shame of another failed relationship likely discouraged her from acknowledging the extent of the abuse she was tolerating. Perhaps she was waiting for her older son to arrive and hoping that his presence would lessen the violence directed at her. Calling attention to her plight might have prevented her son from coming to the United States for college, for which she had long planned and prepared. Perhaps, as is common in the Chinese immigrant community, she did not perceive her husband's verbal abuse and controlling behavior as abusive. It may be that only when her husband's actions became physical that LF perceived her husband as an abuser and felt her well-being was in danger. Like many immigrant women, LF may have lacked knowledge about her rights in the United States, or feared police insensitivity toward immigrant communities. In fact, law enforcement personnel are sometimes "super-sensitive" to cultural issues, and fail to respond in standard ways for fear of being accused of not respecting cultural norms.

Cultural norms about family life are also embedded in Florida statute 765 that regulates end-of-life issues, including the identification of who may serve as a proxy decision maker for an incapacitated individual.[2] The hierarchical list of proxies begins with the judicially appointed guardian of the patient, followed by the patient's spouse. These top choices are followed by an adult child of the patient, or a majority of the adult children; a parent of the patient; the adult sibling of the patient or a majority of one's adult siblings; an adult relative of the patient; a close friend of the patient; or lastly, a licensed clinical social worker. Embedded in this hierarchy is the assumption that one's spouse (since most formally competent adults lack oversight by a guardian) has one's best interests at heart and is in the best position to make medical decisions. For anyone who is a member of a non-traditional or dysfunctional family system, the hazards of this legal structure become immediately apparent. A homosexual couple unable until recently to marry, for example, would not be able to make medical decisions for one another. The case of Terri Schiavo also revealed the vast difficulties of assuming that individuals on this list with close ties to the patient would agree about the appropriate course of action. Terri Schiavo's husband and parents spent 15 years embroiled in litigation about whether or not to discontinue artificial nutrition and hydration after she was diagnosed as being in a permanent vegetative state.[3]

Laws and ethical principles should ideally reinforce one another. The Florida statute aims to support the extension of a patient's autonomy beyond their capacity to make their own decisions by specifying a hierarchical list of proxy decision makers, standards of evidence about a person's medical treatment preferences, and provisions for writing a living will or advance directive. But clearly in this case, both ethics and law failed to protect the rights of LF. She chose not to pursue her legal right to protection from her violent husband, and inadvertently allowed him to remain in a position of power over her even after he was the cause of her incapacity. Ironically, her husband's right to be seen as innocent until proven guilty, and to be named as his wife's medical decision maker, were better protected under the current legal and ethical structure than were LF's rights to autonomy and protection from bodily harm.

Justice was served when WP was arrested, convicted of murder, and sentenced to life in prison, but it came far too late for LF. The law also made victims of her sons—one who remains in foster care and one who returned to China without the benefit of an American university degree, and likely with a good measure of guilt for his inability to protect his mother from fatal harm. It is difficult to fault the hospital personnel who followed the letter of the law, but unfortunately such diligence does not always lead to an ethical outcome. Court-appointed guardians serve a useful and necessary function, but the process is burdensome—law does not move at the pace sometimes required in medicine. The standards of evidence for a criminal investigation are also cumbersome to apply to medicine and to ethics. The law protected WP's rights, but failed to protect the interests of his battered spouse or embattled sons. The ideals of open communication—to ask the pertinent question, to question standard operating procedure, to voice concern about another person's life and liberty—must also be safeguarded and held as sacred.

Notes

[1]The ethical principle of beneficence requires us to do good things for one another, but it does not obligate us to put ourselves in harm's way, unless we have a professional obligation to do so. For example, an average person need not stop to assist a stranded motorist in a dangerous neighborhood, which might reasonably put their well-being at risk; however, a police officer, who has taken an oath to protect the public, must do so.

[2]Florida Statute 765—Health Care Advance Directives. http://www.leg.state.fl.us/statutes/index.cfm?App_mode=Display_Statute&URL=0700-0799/0765/0765.html.

[3]For more information on the Terri Schiavo case, see Caplan, A., McCartney, J. J., & Sisti, D. A. (2006). *The case of Terri Schiavo: Ethics at the end of life.* Amherst, NY: Prometheus Books.

References

Ahmad, Farah, Sarah Riaz, Paula Barata, and Donna E. Stewart. 2001. Patriarchal beliefs and perceptions of abuse among South Asian immigrant women. *Violence Against Women* 10: 262–282.

Das Dasgupta, Shamita. 2000. Charting the course: An overview of domestic violence in the South Asian community in the United States. *Journal of Social Distress and the Homeless* 9: 173–185.

Fadiman, Anne. 1997. *The spirit catches you and you fall down: A Hmong child, her American doctors, and the collision of two cultures.* New York: Farrar, Straus, and Giroux.

Midlarsky, E., A. Venkataramani-Kothari, and M. Plante. 2006. Domestic violence in the Chinese and South Asian immigrant communities. *Annals of the New York Academy of Sciences* 1087: 279–300.

Pew Research Center. 2013. The rise of Asian-Americans. http://www.pewsocialtrends.org/files/2013/04/Asian-Americans-new-full-report-04-2013.pdf.

Veatch, Robert M. 2003. *The basics of bioethics (2nd ed)*, 64–84. Upper Saddle River, NJ: Pearson Education.

Part III

Autonomy and Other Ideals: Balancing Benefits and Burdens

In some end-of-life situations, it is challenging to identify an appropriate decision-maker and equally challenging to develop a process that might best lead to ethically, medically, and legally sound decisions. Religious beliefs, immigration status, addiction, and caregiving are some of the factors that lead to the perplexing cases discussed in this section. Specifically, the cases in this section concern the risks of bloodless surgeries for Jehovah's Witness patients, the obligations hospitals face in caring for indigent patients, appropriate informed consent procedures for patients who are unable to adhere to postsurgical care recommendations, and how caregiving burden in late life can lead to poor outcomes for both spouses. A short summary of the cases in this section is given as follows:

Case 12—Something More Important than Life

A 25-year-old woman, newly engaged to a man who was a lifelong Jehovah's Witness, was a good candidate for heart valve replacement in a hospital's bloodless surgery program. Complications due to the patient's Turner Syndrome caused extreme blood loss, and the patient's fiancé refused to allow a transfusion to take place.

Case 13—Are There Limits on Futile Care for Patients in the U.S. Illegally?

This case examines the complex issues involved in caring for a patient in the U.S. illegally who was involved in a motor vehicle accident. He was diagnosed as being in a persistent vegetative state, and proxy decision-making was complicated by his illegal immigration and marital status, along with his estranged relationship with his mother in Mexico.

Case 14—To Treat … or Not to Treat?

This case involves an unfunded patient with advanced head and neck cancer, complex medical care needs, and no safe plan for discharge. To what extent physicians have an obligation to present all possible treatment options is discussed, along with hospitals' responsibilities to insure a safe discharge plan for indigent patients.

Case 15—A Patient's Right to Treatment (and a Physician's Right to Refuse)

A 55-year-old man had altered mental status and was in septic shock after having some dental work done "in a garage." The patient's continued alcohol and illegal drug use made him a poor candidate for surgery for his antibiotic-resistant infection, and the physicians caring for him struggled to develop an appropriate plan of care.

Case 16—A Depressed Caregiver Neglects His Own Health

A man who had been caring for his disabled wife for over 50 years neglected his own health, and his family needed to make a series of medical and other decisions on behalf of him and his wife, often without sufficient documentation or information.

Case 12—Something More Important Than Life

12

MT was a 25-year-old woman born with Turner Syndrome. Turner Syndrome is the second most common genetic disorder affecting females, with approximately 1 in 2000 live births occurring annually. The most common feature of Turner Syndrome is short stature, with an average adult height of 4 feet, 7 inches. Turner Syndrome is usually diagnosed in early childhood when short stature and frequent ear infections are observed. Other consequences of Turner Syndrome can include risk of ovarian failure, Type II diabetes, and hyperthyroidism. Some individuals have learning disabilities or cognitive difficulties, including problems with visual-spatial tasks, memory impairment and attention deficit disorder; some individuals also experience psychological problems such as low self-esteem or isolation.

MT experienced none of these complications, but she did have an aortic heart valve defect. One third to one half of individuals with Turner Syndrome have bicuspid aortic valves, wherein the major blood vessel from the heart has only two, rather than the normal three, components to the valve regulating blood flow. Complications associated with these heart valve defects can be life-threatening. MT had her first aortic valve replacement when she was 13 years old, and upon her most recent visit to the cardiologist she was told she needed to schedule another valve replacement soon. Her previous surgery had been successful with no resulting complications.

MT had recently become engaged to a young man who was raised as a Jehovah's Witness and who was still very involved in the local Kingdom Hall. MT was raised as a Protestant, but she was becoming interested in her fiancé's religious beliefs, which her future in-laws encouraged. The young couple regularly studied the Bible together, guided by a Bible study publication called, "What Does the Bible *Really* Teach?"[1] The next step for MT, before being baptized into her new faith, would be to preach about Jehovah, going house-to-house with her fiancé and other Witnesses.

Jehovah's Witnesses do not believe in receiving blood products, including transfusions of whole blood, red cells, white cells, platelets, and plasma, which can complicate their medical care. Receiving blood fractions and autologous transfusions of the patient's own blood are designated as matters of individual conscience. Witnesses derive their view from Bible passages such as the following:

- **Acts 15: 29** *"that ye abstain ... from blood ..."*
- **Acts 21: 25** *"... Gentiles ... keep themselves from things offered to idols and from blood ..."*
- **Genesis 9: 4** *"But flesh (meat) with...blood...ye shall not eat"*
- **Leviticus 17: 12–14** *"...No soul of you shall eat blood...whosoever eateth it shall be cut off"*.

Jehovah's Witnesses have interpreted *"eating"* of blood in its most general form to include not accepting *"transfusion of whole blood, packed red blood cells [RBCs], and plasma, as well as white blood cells [WBCs] and platelet administration."*[1] In a Canadian newspaper interview in 2002, L.C. Cotton, then associate director of Jehovah's Witnesses hospital information services said: "We feel that the Bible clearly indicates that blood is sacred and it is not to be used for human consumption. Though it doesn't discuss it in medical terms, Jehovah's Witnesses feel that would preclude the acceptance of it in a blood transfusion" (Harrington 2002, A7).

MT's cardiologist suggested she contact her previous cardiac surgeon, but when he learned of her new religious beliefs, he referred her instead to a local teaching hospital that had instituted a Bloodless Medicine and Surgery Program. "Bloodless" in this context means medical or surgical treatment without the use of stored blood or primary blood components, and the program offered blood conservation services for patients, including Jehovah's Witnesses, who wished to avoid blood transfusions during medical and surgical procedures. Such techniques are used for patients who do not want to be exposed to blood products for a variety of reasons: Religious beliefs, fear of infection, desire to avoid possible exposure to blood-borne diseases and viruses, or because of past allergic reactions or complications from blood or blood products.

MT was carefully screened, and was determined to be a good candidate for minimally invasive aortic heart valve replacement surgery at the Center, and the procedure was scheduled. She was asked to review a consent form that stated:

> I direct that no blood transfusions (whole blood, red cells, white cells, platelets or fresh frozen plasma) are to be given to me **under any circumstances, even if physicians deem such necessary to preserve my life or health**.

MT was advised to complete and sign the form only if she was in full agreement with the statement, as it would direct her health care providers to avoid the use of all blood transfusions in her care no matter the consequences. MT signed the consent form, as well as another form specifically designed for Jehovah's Witness patients releasing their physicians from liability:

> You are hereby notified and instructed that I do not accept any transfusion of blood or primary blood products in my treatment. I will accept non-blood expanders and other forms of alternative management.

I make this medical/religious directive as one of Jehovah's Witnesses. I understand that the attendant physicians may feel that blood transfusions or primary components are necessary. I do not share their opinion and adhere to the instructions given in this notice and in the Bible's command such as: "*Keep abstaining … from blood*" (Acts 15: 28, 29).

I have carefully considered this matter, and my instructions are not going to change because I am unconscious. I release physicians, anesthesiologists, hospitals and their personnel from liability for any damages that may be caused by my refusal of blood, despite their otherwise competent care.

MT signed this release form, and she also initialed another form documenting her refusal to accept blood fractions which contain proteins extracted from plasma, such as erythropoietin, neupogen, human derived clotting factor concentrates, immune globulins, and albumin. She indicated her acceptance of the use of a heart-lung machine, which her cardiac surgeon had told her would be necessary, but she indicated her refusal of other procedures or equipment that would use her own blood, including dialysis, blood salvage, hemodilution, or plasmapheresis.[2]

On the day of her surgery MT was accompanied by her fiancé, her future mother-in-law, and by an Elder from the local Jehovah's Witness Kingdom Hall. Her preferences regarding blood transfusions were documented in her medical record, and she wore a wristband into surgery stating "No Blood." A "No Blood" sticker was on the front of her medical chart, and a similar sign was affixed to the door of her hospital room.

The surgical procedure started routinely, but complications due to her Turner Syndrome developed midway through the surgery, and her blood loss was considerable. The surgical team became concerned that MT would die without the administration of a small amount of transfused blood. The social worker at the Bloodless Surgery Center was experienced in working with Jehovah's Witness patients and families, and she prepared herself to talk with MT's fiancé to present the certain benefit that would come from allowing a small transfusion. She was also prepared to speak candidly about the deadly consequences of refusing to give consent for the transfusion. The social worker's carefully prepared arguments failed to persuade MT's representatives, and she returned to the operating room to report the news to the surgical team. "The patient's family members just keep saying no! And the Elder from the Kingdom Hall keeps reminding me of all the 'no blood' documentation that MT signed," she explained. MT's condition had continued to worsen, and the cardiac surgeon, who had never lost a patient in surgery, asked the social worker to try again. She did so, saying, "MT will likely die without a blood transfusion!" Again, MT's fiancé, his mother and the Church Elder remained steadfast: MT's documented preferences refusing blood products were to be honored, the expected consequences notwithstanding.

The social worker tearfully returned to the surgical suite with this news, and again, the surgeon urged her to try one more time. She did so, explaining this time that MT was now on the brink of death. The Church Elder spoke on behalf of MT and said, "MT clearly refused to consent to receive any blood products. She signed every document you provided, and now you are obligated to follow her wishes." The social worker carefully explained that MT was sure to die without their consent, and, again, all speaking on her behalf refused, stating MT was aware of this potential outcome and that they were there to insure her wishes were followed. When the social worker once again conveyed this news to the surgical team, there appeared to be no option other than to allow MT to die. MT was pronounced dead in the operating room. The surgeon then met with the family to explain what he saw as the preventable circumstances surrounding MT's death.

Discussion Questions

1. Is there clear evidence to support the idea that MT's beliefs were in line with those of Jehovah's Witnesses?
2. Would the surgeon have been justified in giving MT the transfused blood, and not informing the family that he had done so to save her life?
3. Should the surgeon have attempted to secure an emergency guardianship for MT to override the decision-making authority she had implicitly granted to her fiancé? Why or why not?

A Bioethicist Responds

This case is troubling in several regards, beginning with the general description of the patient. The case narrative presents MT as one might almost expect a textbook Turner Syndrome patient to appear, that is, a fairly typical one with no major problems, save the heart defect.[3] This would suggest that MT should have had no significant difficulties with regard to intelligence, cognition, psychological or social adjustment, allowing only for the possibilities of a learning disability in math, or difficulty in social situations, such as understanding other people's emotions or reactions. Indeed, nothing is reported here to signal that she might have had any learning disabilities or difficulties in social situations, but we simply do not have a complete history, as is often the nature with case reviews. Nevertheless, given that difficulty in social situations may not be uncommon in patients with Turner Syndrome during their teen-age and young adult years, there is reason for caution here.

A second concern has to do with the consent form that MT initialed documenting her refusal to accept blood fractions containing proteins extracted from plasma, such as erythropoietin. Jehovah's Witnesses are afforded the freedom by their faith to accept or refuse the use of these blood fractions at their discretion.[4] The cause for concern, however, comes from the fact that while MT chose the more conservative route of refusing not only erythropoietin, but also mechanical procedures which would use her own blood, such as dialysis, blood salvage, hemodilution and plasmapheresis, she consented to the use of a heart-lung machine, another choice also left to the discretion of Jehovah's Witness patients. There would appear to be an apparent inconsistency in reasoning here, at least on the surface.

A final concern centers on the roles of MT's fiancé, her future mother-in-law and the Elder of the local Kingdom Hall, and this has to do with the influence these individuals had on MT individually as well as collectively. The three foregoing concerns will be taken up sequentially.

As noted above, case reviews provide limited information about a patient, and it is always tempting to wish more than what appears to be the barest of essentials had been given. For example, despite the fact that MT had apparently never manifested any of the problems or complications that can be associated with Turner Syndrome, save the obvious heart valve defect, we know nothing of her social background aside from her current relationship to her fiancé. We do not know how much dating experience she has had prior to this, if any; we do not know what her concept of herself, or her own self-image, may be like; we do not know whether she might be particularly self-assured, timid, easily impressed or swayed by others, or anything else about her as a person. And, given that she was apparently on her own, without her own family in the picture to help or support her in any regard, it might be altogether reasonable to conclude that her fiancé, his mother and their religious community, ready and eager to provide her with a highly-structured system of values and all-encompassing approach to life on multiple levels, may well answer her need for connection, attachment, a sense of belonging and, ultimately, some meaning in her life. It would behoove the observer of any young woman this age, therefore, to ask what kind and degree of influence other parties might have on her decision making, but the fact that MT was a Turner Syndrome patient should warrant even greater concern. It is not unreasonable to wonder whether or not, and to what extent, MT may have been pressured into wanting to become a Jehovah's Witness, and for what reason(s). Could it have been that this was the first young woman her fiancé's mother had met and approved, and that this well-meaning mother was merely trying to encourage a good marriage for her son? Or, did the son also have some sort of disability to which we are not privy, and was the mother, therefore, desperately and cunningly pressuring MT into converting so as to be marriageable? This concern could well be overplayed, but reason for caution remains regarding the possibility of coercion.

The second concern is the most puzzling of the three. The religious position of Jehovah's Witnesses with regard to whole blood transfusions is clear and unambiguous; it was well laid out and explained years ago in *The Watch Tower* by Professor Frank H. Gorman.[4] The Jehovah's Witnesses' position on the four primary components of whole blood is also clear where red cells, white cells, platelets and plasma are all forbidden to be transferred from one person to another. But, it is also clear in publications of the Watch Tower Bible and Tract Society of Pennsylvania that there is no absolute prohibition about the use of blood fractions, where one is dealing with the building blocks of the primary components of whole blood, for example, such things as the hemoglobin of red cells, clotting factors, antibodies and gamma globulin derived from plasma, and the interferons or interleukins derived from white cells. It is argued that inasmuch as a committed Jehovah's Witness "may reason that at some point fractions that have been extracted from blood cease to represent the life of the creature from which the blood was taken," each individual "must make his or her own conscientious decision before God."[4] In other words, Jehovah's Witnesses may, at their discretion and after prayerful consideration before God, choose to accept the administration of blood fractions, or not. Thus, if one reads carefully the materials prepared for Witnesses themselves, this is a matter left entirely up to them and their Lord. Likewise, it is also clear in Witness literature that normal saline, Ringer's solution and dextran can be used as non-blood plasma expanders in order to maintain fluid volume within the circulatory system.[4] Furthermore, procedures whereby a patient's blood is temporarily removed from the body, though not "stored" at any distance nor for any significant time apart from the body, may be permitted provided the patient's conscience will allow her/him still to view the blood as an integral part of the self and thus not requiring that it be poured "out on the ground" (Deuteronomy 12: 23, 24). This would include such things as hemodilution, cell salvage and the use of a heart-lung bypass machine.

Thus, the consents that MT signed and initialed on the day of surgery appear rather puzzling. It was entirely within her discretion, according to Jehovah's Witness teachings, for her to refuse the use of blood fractions or any procedures or equipment that would use her own blood, but she did consent to the use of a heart-lung machine while excluding all other matters of discretion. The case report makes clear that the surgeon had told her the heart-lung assist would be necessary during the surgery; perhaps it may be presumed that the influence of the surgeon was enough to persuade her that this was permissible, particularly since this procedure fell into that class of issues to be decided upon in the discerning mind of the believer. There is no other information available that would explain MT's choice of the heart-lung machine, her exclusion of other discretionary choices, nor the reasons for her choice.

This, then, brings up the third concern, the influence others had on MT individually and collectively. Doubtless the surgeon would not have agreed to operate, absent consent for the use of the heart-lung machine, a perfectly understandable requirement. MT was given a simple choice in this matter, and she chose the prudent course, presumably based upon her understanding of Jehovah's Witness teachings that in doing so she was exercising conscientious discretion in keeping with the faith. She seemed to be in control of things, at least at this juncture. Yet one cannot help wonder about who was in control at other key moments of this story. Her fiancé's mother was noted to have encouraged MT's interest in their family's religious beliefs, and she certainly spoke on MT's behalf, along with her son, in refusing a transfusion or blood products during the surgery. The Church Elder was also steadfast in refusing any of these life-saving measures.

Questions remain in light of these three concerns. To what extent was MT coerced into accepting the teachings of Jehovah's Witnesses? Since she had apparently not yet been baptized, could she really be said to be a member of the faith? Even if she wished to follow Jehovah's Witnesses teachings, what right did her fiancé, his mother and the Church Elder have to speak so forcefully for her? There is no clear evidence that she ever formally assigned surrogacy or durable power of attorney to anyone. It is quite understandable that with the imposing presence of the fiancé, his mother and the Church Elder in the waiting room, the surgeon would send the social worker to them three times in attempts to secure an override of his patient's pre-operative consent stipulating her refusal of blood. It might be argued that even had MT not officially named one of these three persons to speak for her, and, again, we are not sure, the surgeon could presumably have sought emergency guardianship. Pursuing this route could have been very problematic, however, particularly as it would in no way be entirely up to the surgeon. The hospital's risk management office would be responsible for seeking the court order, but even before doing so that office would undoubtedly seriously weigh the almost certain legal fight resulting from such action; the hospital could well decline to support the surgeon's request for guardianship. And given the emergency situation requiring immediate action, even the most expeditious guardianship process would likely have been too late.

Still, one might also ask if a little sleight-of-hand and manipulated reasoning could not have been employed by the surgeon in order to achieve the positive outcome he sought. That is, he might have been able to administer the needed transfusion to save MT's life, never reveal this to her nor her three companions, and reason to himself that while he had indeed not respected her wishes, the harm done to her would have been far outweighed by the good end result, knowing all along that had he not done so he would have allowed her to die needlessly. Concomitantly, he might perhaps have reasoned that in never revealing this fact she would be done no harm whatsoever, especially if she were not yet a baptized Jehovah's Witness. And, of course, such an action would allow him to maintain his record as one "who had never lost a patient in surgery."

This surgeon did exactly the right thing, however: He completely ignored his own concerns; he upheld the rights of his patient, fully respecting her wishes as stated in the consent that she had signed, puzzling as those stipulations may appear on paper and in the case report; and, he correctly judged that she was a competent adult when she signed the consent. Regrettable as it was for him to allow his patient to expire on the operating table, he made the only ethical choice that he could. And, he arrived at his choice through an exercise of prudence, using wisdom in carefully balancing principles with virtues to arrive at the only possible good, given the constraints placed upon him at the time.

The problem remains, however, that while MT may have been medically competent enough to consent to the procedure that she did, she should still be viewed as a vulnerable person by virtue of her diagnosis as a Turner Syndrome patient. And it is in that context that she can and should be seen as one who has most probably been unduly influenced by her fiancé and his family into accepting the tenets of their faith even before being formally accepted or admitted into that faith. Moreover, what at one point appears merely as influence, is later more suggestive of coercion as when MT reached the point of requiring heart surgery and was accompanied to the hospital by a Church Elder representing the faith, ready to remind her and the surgeon of certain obligations. The ethical issue to be resolved in this case is not one that can be resolved at the eleventh hour when the patient is in the OR and needing blood; by then it is too late. The ethical issue here is one to be resolved much earlier, when a patient with Turner Syndrome, or any patient in a vulnerable position, should be given careful assistance and preparation in advance of the consenting process so that all concerned can be as secure as possible in the knowledge that the ultimate consent will be as truly informed and unfettered as possible.

A Health Communication Scholar Responds

A foundational operationalization of our ethical dictum to have respect for persons and to honor autonomy is the right of a patient to refuse any medical treatment for any reason, even if doing so will end his or her life. Doctors and other health care professionals, as well as family members, are generally more comfortable when a person who has chosen to forego life-support has an advance directive, has lived a long life, has tried aggressive medical treatment, has a documented religious objection to certain treatments, and if his or her stated treatment preferences appeared consistent with other decisions and life choices (Roscoe and Tullis 2015; Schenck and Roscoe 2009). Perceived agreement between a person's previous life choices and end-of-life treatment preferences is not strictly necessary, but it provides comfort to those who witness the implementation of a person's treatment refusal and subsequent death.

An initial point of inquiry in this case is to examine the extent to which we believe MT truly believed and felt committed to her relatively new, and supremely consequential, Jehovah's Witness beliefs. Her conversion to this new faith tradition literally cost her life. We know from the case description that MT's interest in becoming a practicing Jehovah's Witness was a consequence of her engagement to a young man who was a Jehovah's Witness. They were newly engaged, and we do not know how long she knew her fiancé prior to their decision to marry, how they met, or the quality of their relationship. We know also that MT's conversion to the Jehovah's Witness faith tradition and values were important to her young man and to his family, and that they encouraged and supported MT as she began to live as a Witness. MT was not baptized at the time of her surgery, but she was moving in that direction; having a specific experience such as baptism may be important to one's membership in a religious organization, but is not required as proof of one's evolving beliefs or commitment to a value system that might influence one's end-of-life treatment preferences.

We know too that MT sought the referral to a bloodless surgery center, and that as far as we know signed the consent forms of her own volition. Was she coerced by her boyfriend, his family, or by the Church Elders? We have no evidence of this. Would she have signed the forms if her own Protestant family members were present? We do not know this either, nor do we know the nature of MT's relationship with her family of origin or the depth of her past engagement with Protestant teachings. We know only that MT was accompanied by her fiancé, his mother and the Church Elder, that she signed all necessary consent forms indicating her refusal of blood products, and that she was perceived to be competent to make these decisions and willing to have her new Jehovah's Witness family speak on her behalf.

When making moral judgments, we tend to prefer consistency in another's belief systems; both authors of this casebook have previously written that one aspect that might define "a good death," although that remains a problematic concept, is that a person's death is consistent with choices they had made during their lives (Schenck and Roscoe 2009). But people change, sometimes radically, and often in response to illness, love, or other transformative experiences. While we might respect a person who remains true to their nature or steadfast in their beliefs, practically speaking we honor the autonomy of one another as individuals by honoring the most recent iteration of the ways in which we present ourselves. For instance, a person's advance directive may always be overturned by a more recent version, or by a patient's ability to change it verbally in response to new information. While MT's beliefs and desire to belong to the Jehovah's Witness community may seem sudden or new to us, we have no choice but to honor these commitments given the ways in which her clear choices were communicated and documented.

It is tragic when a young person dies, especially under circumstances that appear preventable. In MT's case, her fiancé was not presented with a decision to allow a procedure with uncertain benefit, or one that would impose physical burdens or undue suffering. Quite the opposite: A simple transfusion of blood would allow MT to survive her surgery, and go on to marry and begin her new life in improved health. But the choice presented to allow the blood transfusion was costly and imposed considerable burden in spiritual terms. If her fiancé had allowed the transfusion against MT's expressly documented preferences, what would this communicate to her about his trustworthiness or the steadfastness of his commitment to her or to his faith? If her fiancé allowed the transfusion and MT was grateful for his life-saving decision, she (and they) would have to live with the consequences of being unwelcome as part of her in-law's family, and potential pariahs in their chosen faith community. It is sometimes noted that the most recent convert is the most devout—one has only to witness the disgust on the face of someone who has recently quit smoking when someone nearby lights up. Living with the burden of having transgressed this most fundamental tenet of the Jehovah's Witness faith tradition may well have been more than MT could bear. As a very wise person said during an ethics committee discussion of a similar case, "there are worse fates than dying for one's faith."

We know the consequences of MT's treatment preferences for her—she died because of her refusal to accept blood products. Did she die for love, or for faith? Perhaps both. We can only imagine the depths of grief experienced by her fiancé and his family, especially since they were present and in charge of making sure MT's preferences were honored and were witnesses to the deadly consequences of those choices. Instead of celebrating the success of her surgery and recent engagement, they must turn to the much grimmer tasks at hand: to notify her next-of-kin, who may not even know of their existence, and to undertake plans for her burial.

The cardiac surgeon, who just lost his first patient in surgery, the other members of the surgical team, the administrators of the bloodless medical and surgery center, and the social worker who was caught in the middle between a family with no intention of altering their religious beliefs or amending the documented preferences of MT are all parties to this grief as well. It is a truism of modern medical care in the U.S. that "no one dies in the OR," and there can be no doubt that this case was deeply felt by the surgical team. The cardiac surgeon might well question his commitment to the bloodless surgery program, possibly leaving a large void in the expertise that could be offered to other patients. Did the surgeon contemplate transfusing the blood and just not informing MT or her family? Her life would have been saved, but surely one cannot condone a return to the kind of strong paternalism that permits physicians to act in ways that expressly contradict their patients' stated preferences.

Can we imagine other communicative tactics the social worker might have tried? Are strategies such as manipulation, strong fear appeals, or outright deception ever ethically permissible? It appears from the case description that the surgical team had identified the easiest and most effective intervention—a blood transfusion—and that MT had clearly documented her refusal of any interventions save the heart-lung machine. Should the social worker have tried harder to frighten or shock the family? It appears that she was straightforward about the consequences of continued refusal of blood products; MT would die. And in any case, fear appeals are limited in their persuasive power (Witte 1992). There is some evidence that strong fear appeals might be most effective, but overall, the literature on the efficacy of fear appeals is inconsistent (Witte and Allen 2000). And faith might be the strongest antidote for fear in any case.

Notes

[1]"What does the Bible *Really* Teach?" (http://www.jw.org/en/publications/books/bible-teach/).

[2]*Heart-lung machines* take over the functions of the patient's heart and lungs during surgery to maintain blood circulation and oxygenation of the body. *Dialysis* is a procedure that circulates a patient's blood through a machine that filters and cleans it before returning it to the patient. *Blood salvage* is a process that involves recovering blood lost during surgery and reinfusing it into the patient. *Hemodilution* is a medical procedure to reduce the number of red blood cells lost during surgery; it involves the pre-operative withdrawal of one or more units of whole blood, immediate replacement with IV fluid, and post-operation reinfusion of the withdrawn blood. *Plasmapheresis* is the removal, treatment, and replacement of blood plasma or a plasma substitute.

[3]For more information, see National Institutes of Health, National Human Genome Research Institute—Turner Syndrome; www.genome.gov/19519119; and Mayo Foundation for Medical Education and Research; www.mayoclinic.org/diseases-conditions/turner-syndrome/basics/symptoms/con-20032572.

[4]The following publications of the Watch Tower Bible and Tract Society of Pennsylvania informed the information presented in the case analysis:
Keep yourself in God's love. 2015; The Watch Tower 2000; Reasoning from the Scriptures 1989.

References

Harrington, Carol. 2002. Father shunned by family for defying faith to save child. *Toronto Star* A7.

Roscoe, Lori A., and Jillian A. Tullis. 2015. The meaning of everything: Communication at the end of life. *Journal of Medicine and the Person* 13: 75–81. https://doi.org/10.1007/s12682-015-0205-x.

Schenck, David P., and Lori A. Roscoe. 2009. In search of a good death. *Journal of Medical Humanities* 30: 61–72.

Witte, Kim. 1992. Putting the fear back into fear appeals: The extended parallel process model. *Communication Monographs* 59: 329–349. https://doi.org/10.1080/03637759209376276.

Witte, Kim, and M. Allen. 2000. A meta-analysis of fear appeals: Implications for public health campaigns. *Health Education & Behavior* 27: 591–615. https://doi.org/10.1177/1090198 10002700506.

Case 13—Are There Limits on Futile Care for Patients in the U.S. Illegally?

13

JG was a 39-year-old Mexican man who was involved in a serious motor vehicle accident in Florida on New Year's Eve. He apparently lost control of his car and slammed into a highway median divider at high speed. JG was not wearing a seatbelt and was thrown from the car, and his injuries were extensive. In addition to abrasions, bruises, and broken bones, he also suffered significant head trauma. JG was taken by ambulance to the Emergency Department of the nearest hospital. He was taken to surgery immediately to repair his shattered pelvis, shoulder and right leg, and he was then admitted to the Intensive Care Unit (ICU) unconscious and breathing with the support of a ventilator.

JG had been residing in the U.S. for approximately 15 years but was undocumented. He was employed in a machine shop and lived with his significant other. The couple was not married and had two children together, ages 8 and 10. JG did not have an advance directive and had not appointed a health care surrogate. His only biological relative living nearby was a male cousin of approximately the same age. The two men were very close and saw one another frequently. The cousin was contacted and he agreed to serve as JG's proxy decision maker. He wanted all available medical treatment to be given to his cousin while his condition and chances for recovery were being determined. Both JG's cousin and partner were concerned about him, and visited frequently in the ensuing days and weeks.

Because JG was living in the U.S. illegally, he had no insurance coverage nor was he eligible for any governmental aid. *Undocumented immigrants* are not eligible to buy health insurance coverage through the Affordable Care Act (ACA), nor are they eligible for premium tax credits or other savings on ACA Marketplace plans. Medicare and Medicaid are also out of the question for *undocumented immigrants*. JG's cousin and significant other lacked the financial resources to pay for his care, and the cost was prohibitive.[1] After 2 weeks, a tracheostomy was performed (for which JG's cousin provided consent) so ventilator support could continue longer term. After 6 weeks in the ICU, JG had transitioned from a coma into what his physicians determined was a persistent vegetative state.[2] JG had brief periods of waking and sleeping, and occasionally moaned or sighed. He required

© Springer International Publishing AG 2017
L. A. Roscoe and D. P. Schenck, *Communication and Bioethics at the End of Life*,
https://doi.org/10.1007/978-3-319-70920-8_13

extensive care; in addition to being ventilator dependent and on a feeding tube, which both required their own care routines, he was catheterized, needed to be turned to prevent pressure ulcers, and had to be treated with antibiotics for a persistent urinary tract infection as well as with anti-seizure medications. JG's cousin and partner were both informed there was no hope for JG to regain a meaningful level of consciousness, and that continuing to provide such intensive care was futile. When discussing options, the hospital's discharge planner said it was unlikely that a nursing home or sub-acute rehabilitation facility would accept JG as a resident because of his undocumented and uninsured status as well as the extent of his care requirements (JG could not be successfully weaned from the ventilator). JG's attending physician explained that it was possible for his cousin to make a decision to stop providing ventilator support, and that JG would be kept comfortable during the process and that his family could be with him as he died. The cousin and JG's partner tearfully discussed what to do, and eventually agreed that removing the ventilator and allowing JG to die peacefully was a sad but necessary decision to make. No immediate date was set for the terminal extubation, but all agreed it should happen soon.

In the meantime, the social worker assigned to JG's floor had discovered that he had a mother who lived in a small town in Mexico, from whom he was estranged. The mother and son had not spoken for over 15 years. A call was placed to JG's mother and her son's dire condition was explained to her. The nature of her son's injuries, the futility of continuing treatment, and the fact that JG would be kept comfortable during the process of discontinuing treatment were all explained in detail, carefully translated from English to Spanish and back again. The hospital administrators had also agreed to pay for JG's body to be embalmed by a local funeral home and transported back to Mexico, with additional funds for a funeral for JG. It was also explained that the mother and her family would not be required to reimburse the hospital, the funeral home, or the airline for any costs associated with JG's care and after-death expenses. After a long conversation, JG's mother tearfully agreed with the plan that had been put in place by the cousin—JG's life supportive care would be removed and he would be allowed to die. JG's mother then asked, "¿Puedo ver a mi hijo una vez mas?" (Can I see my son one more time?) The social worker was able to arrange for a Skype session, of which JG's cousin was unaware. The Skype session was to occur in three days, and the day after JG's cousin and his partner were to come to the hospital to be with JG as he was removed from life support and allowed to die. The funeral home and airline were ready to make their contributions thereafter.

The Skype session took place, and upon seeing her son for the first time in many years, JG's mother pleaded, "¡Usted debe hacer todo para mantenerlo vivo!" (You must do everything to keep him alive!). The nurses caring for JG thought that seeing her son being suctioned would convince her that he was suffering, but instead JG's mother decided that it was the suctioning that caused her son's suffering, not the life support, and she insisted that all life support and other medical care continue indefinitely.[3] She also reiterated the fact that she did not have the financial resources to pay for her son's care. The social worker explained that the hospital was not equipped to provide such care for JG indefinitely, and that an alternative placement for him would need to be determined. JG's mother stated that no appropriate care facilities existed in Mexico and that it would be impossible for her to care for JG in her home.

During his next visit, JG's cousin was told about the Skype session, and he was furious. "JG and his mother haven't spoken in years! I never would have agreed to this plan," he said angrily. JG's cousin was also angry that his decision-making authority had been usurped by the patient's mother, as he felt that he knew JG far better at this point than did his mother, and he was certain JG would prefer for him to be making decisions on his behalf. He stopped coming to the hospital for regular visits, and told the social worker she would now have to deal with the patient's mother instead of with him. He also repeated that neither he nor JG's partner had the financial resources to pay for his continued care, and neither had the space, time, or ability to care for him in their homes. JG's partner was bewildered by this sudden change of events. Both she and the cousin were confused about why JG's mother had been contacted as a proxy decision maker for him, since previously the physicians had been willing to consult with the cousin about medical decisions for JG.

The social worker and discharge planner assigned to JG's case called every possible nursing home facility in the state of Florida, and were unable to locate a facility that would accept JG as a resident. JG continued to receive care in the hospital, much to the chagrin of the nurses and physicians charged with his care, which they believed to be futile and not in his best interest. As of this writing, JG continued to receive care in the hospital, where he has been in a persistent vegetative state for over 6 months. His prognosis for recovery is nonexistent, but his life expectancy, if infections are treated and he remains on life support, is at least several years.[4]

Discussion Questions

1. Who should make medical decisions for incapacitated patients who have not named a surrogate?
2. How are hospitals to provide compassionate care to undocumented and/or uninsured patients while still being good stewards of their financial and other resources (beds, expertise, etc.)?

A Bioethicist Responds

This is another very sad, unfortunate case made all the more disturbing because of the combined social, familial, interpersonal and legal issues that would appear to block an acceptable resolution to the situation in which JG has been left. At first glance, it would appear as if the hospital is faced with having to absorb the long-term costs of his care, which could soon run into the millions of dollars. How, then might this problem be resolved, and what are the ethical issues involved in seeking resolution?

The first major issue to be addressed should be that of withdrawal of the ventilator. As mentioned in the case narrative above, once JG's cousin and partner had discussed his situation with the attending physician, the two former individuals made the only decision they felt proper, which was to allow removal of the ventilator, thus allowing JG to die peacefully. Under normal circumstances, of course, such care and treatment are not simply, or summarily, withdrawn by attending physicians from their hospitalized patients unless patients or their legally appointed representatives have signed consents to this effect, and/or valid advance directives specifying the circumstances under which withdrawal is to be carried out. It is not infrequently the case, however, that in situations of medically incompetent patients, it may little matter what standard practice, a reasonable and justifiable decision of a legally appointed representative, or a clear, unambiguous and properly notarized advance directive may be. Family members, and even close friends, can have widely different opinions about what the patient may have wanted, but physicians and hospitals are not wont to find themselves between factions or facing threatened legal action, even if they may have the "right" on their side.

In point of fact, JG's physician and the hospital do have the right on their side in this case, and they would have been entirely justified in withdrawing the ventilator and allowing him to die in a humane manner as initially proposed and agreed to by all prior to the unfortunate Skype session. Yet nobody able to authorize such a decision would likely do so, for virtually no responsible party, most especially the hospital, would lightly consider the negative publicity that such action would most surely generate. "Care Withdrawn; Illegal Mexican Dies," or some similar headline, is about all that would be necessary to give rise to a scandalous tale about a nameless, faceless, expendable, undocumented person who had entered this country illegally, had enjoyed an American way of life, had perhaps not paid his fair share of taxes, had run up enormous medical bills he was unable to pay, and who, because he was poor, undocumented and needing costly intensive care for an indefinite future, had all his care withdrawn so that he died. Add in reference to his Mexican heritage and the story takes on even greater proportions, while the media volume increases, perhaps, as well.

This case occurred during the extraordinarily volatile 2016 U.S. Presidential Campaign, a time when numerous views on immigration and certain immigrant populations were expressed by different sides; therefore, the facts of JG's national origin and undocumented status can hardly go unnoted here. This is in no way meant to suggest that politics should play a part in bioethical decision-making. In this writer's view, it should not, but since it would be impossible to ignore the facts of this case, the temporal-social context during which it occurred, and issues likely to surface in discussion as a matter of consequence, JG's undocumented status will be addressed. A review of Mexican immigration data will be useful as a first step.

The latest U.S. Census Bureau's American Community Survey (ACS) reports there were over 11.7 million immigrants from Mexico residing in the United States in 2014, thereby accounting for twenty-eight percent (28%) of all U.S. immigrants. The majority of these Mexican immigrants were concentrated in the West and Southwest, with only two percent (2%) estimated to have been living in Florida.

This may appear to be a small number, but considering that the total population of Florida in 2014 was 19,893,297, and that 3,973,515 of them were foreign born, immigrants comprised twenty percent (20%) of the state's population. The official number of foreign- born Mexicans in Florida in 2014 was 282,594.[5] It is further noted that Mexicans represent the largest unauthorized group of foreign-born persons in the U.S., not surprisingly perhaps because of the relatively porous border between the U.S. and Mexico. And of the 610,000 total undocumented persons (from all foreign countries/regions of birth) living in Florida in 2014, 171,000 of them, or twenty-eight percent (28%), were estimated to have come from Mexico.[6] Comparing this figure with that of the total number of foreign-born Mexicans for 2014 (282,594) would mean that approximately sixty-one percent (61%) were undocumented.

The presence of undocumented Mexicans in Florida has been a fact of life for many decades. It is widely believed that the harvesting of fruit and vegetable crops at current market prices simply would not likely occur but for migrant farm workers, most of whom are from Mexico. That more than half of them are estimated to be undocumented is an issue about which direct action appears to have been "deferred" or "suspended" by authorities in the absence of major problems such as crimes against persons, trafficking and blatant illegal attempts at border entry, for the simple reason that Hispanic migrant farm workers are willing to perform labor for wages, often under less than desirable working and living conditions and without recourse, than would most American citizens. Yet throughout the history of immigration into the United States, persons from Mexico have suffered prejudice and discrimination, being viewed mainly as a ready pool of low-paid laborers, as less intelligent than White Americans, as racially inferior, and even as sources of disease.

Natalia Molina's review of American attitudes toward the increasing Mexican population in the United States during the early and middle years of the 20th century offers critical insights into issues surrounding immigrants today (2011). As she points out, immigration laws did not severely restrict Mexican immigration during the very early years of the last century. In fact, restrictive laws did not begin until 1917 with the Immigration Act. However, medical screenings had begun at the U.S.-Mexico border even before that because of concerns to limit admission to a biologically fit working class. Health concerns were coming to a head in 1916 when it was generally believed that Mexicans were bringing disease into the United States. The issue reached its peak in the summer of 1916 with a serious outbreak of typhus in California. U.S. health officials saw Mexicans as the unique carriers of this deadly disease, thereby making race the organizing principle for understanding typhus, which in turn added a medicalized dimension to the nativist views of those who already saw Mexicans as racially inferior to Whites. This linkage between Mexican immigration and disease became further solidified in 1942 in what was known as the Bracero Program, a program jointly operated by the U.S. and Mexican governments by which Mexican workers would enter the U.S. to work in agriculture, on railroads or in other industries. Mexican workers were given rigorous health and psychological exams by both governments prior to beginning work in the U.S., but once settled in their labor camps, conditions there ultimately and ironically led to poor health among this migrant population, and Mexicans became even further stigmatized as bearers of disease. The Bracero Program was discontinued in 1964. In any case, the next step toward seeing Mexicans as undesirables did not have to be a big one.

Molina then convincingly demonstrates that the American view of Mexican immigrants became "medicalized" through association with disease, and they were then deemed undesirable. This then led to their becoming victims of racism, despite evidence supporting the fact that the diseases for which they were blamed were not inherent in them but due to their living conditions once settled in the U.S. No matter. Their association with disease would seem to hold within itself the inevitable foregone conclusion, for she states, "The consistent representation of Mexicans as disease carriers unworthy of social membership in US society led to the conclusion that they were unworthy recipients of publicly funded health care" (Molina 2011, 1029). And this, she affirms, has developed into the current practice found in some hospitals that have repatriated unfunded, undocumented patients requiring long-term care, a practice that serves to justify the belief that these persons are unworthy.

JG was not being cared for in one of those hospitals referred to above that have been repatriating undocumented persons needing long-term care, and the issue here is not to deal with that more global problem. Yet as was suggested earlier, JG's background combined with the socio-politics of the time during which the problems in this case occurred could not have failed to register as factors to consider, however briefly, in the minds of those attempting resolution to his situation. It would be impossible to ignore the obvious: that he was an undocumented person from Mexico; that he had lived in this country for fifteen years; that the image "Mexican" conveys a particular stereotyped image to the American mind; that many Americans hold racist views of Mexicans; and, that since Americans often believe Mexicans to be undocumented, they have no right to the same health care as American citizens. It would be a mistake, however, to fall into the trap of attempting to sort out any of these issues either individually or in groups. The dilemma of what to do regarding JG will never be resolved by approaching it from social, political or legal points of view. In this writer's view, the only fair and humane way to approach resolution to this issue is by asking once again the prudential question: "What is the right thing to do for this patient?"

Identifying the right thing to do in this case will emphasize actions that do good, do little or no harm and are fair. It would be difficult to say what could be done that would do good for JG at this point since his physician has determined there to be no hope for improvement. Surely there are those who would argue that the most good to be done would be to maintain him as he is until his death from natural causes; others would assert that to do so would provide him with none of the goods of life. Inversely, it could be argued that maintaining him in his present condition indefinitely would not only not be doing him any good, but would in fact be doing harm, forcing him to remain alive with no apparent cognitive functions or ability to enjoy being alive.

One could take the position that since JG's cousin no longer seems willing to accept responsibility as health care surrogate, and that since JG's mother has no legal standing in the United States to act as his legal surrogate, as she presumably could were he a citizen of this country, his physician and the hospital could decide on the basis of futility to remove the ventilator and allow him to die. This could well be seen as the most beneficent act possible, insofar as continued maintenance and treatment have been judged to be futile. It fairly goes without saying, however, that there will be detractors to this view who would insist that this does nothing but harm in that it causes his death.

The issue of justice or fairness must also be considered not only on the level of what is fair to the individual patient but what is fair to society. A decision to withdraw treatment would ensure greatest fairness, in terms of resource utilization to the hospital, although the institution would not want to proceed without first making every effort to work with the patient's mother, on a personal level, to help her understand how this could be the right and good thing to do. This could mean sending a special envoy to the mother's house in the person of a Spanish-speaking counselor, priest, physician or whoever might be viewed as the kind of person capable of communicating with her. It could go a long way to avoiding the negative publicity almost sure to result if the way were not paved ahead of time.

Another solution to the hospital's dilemma should be a thorough investigation of skilled nursing facilities in Mexico. It has only been reported that JG's mother has stated that "no appropriate care facilities existed in Mexico," but there is no indication that the hospital social worker has done the necessary background work to determine the actual situation in Mexico and whether a successful transfer could be arranged.

In any event, despite all the efforts that have already gone into trying to resolve this very difficult issue, questions remain unanswered that should be addressed before it is assumed that nothing is to be done other than expect the hospital to care for JG in his current state for the unforeseeable future. If, in the end, all efforts fail in attempting to work with the mother and hospital to their mutual satisfaction, the hospital should feel comfortable in discontinuing life-support.

A Health Communication Scholar Responds

We include this case in our "end-of-life" casebook even though as of this writing JG was still alive and residing in the hospital that provided his original trauma care. Patients in a persistent vegetative state present a complicated array of end-of-life considerations. Under Florida law, and the laws governing end-of-life care and decision making in many other states, "persistent vegetative state" is explicitly included along with "end-stage conditions" and "terminal diagnoses" as conditions that trigger certain provisions in the law to take effect, such as using advance directives to guide decision-making. A persistent vegetative state is a difficult condition for many families of such patients to come to terms with: Their loved one looks very much alive, and moves and even vocalizes in ways that sometimes appear to be responsive to external stimuli. The brain damage that causes such a state of consciousness is too extensive to make a meaningful response possible, but when one is looking for signs of awareness, such random movements and vocalizations can be imbued with meaning and interpreted as signs of hope for recovery.

The case of Terri Schiavo, the young woman in a persistent vegetative state who lived for 15 years with the assistance of a feeding tube while her husband and parents fought about what they perceived to be her treatment preferences in court, brought these issues into high relief (Roscoe et al. 2006). The Terri Schiavo case was also the first case able to use the power of communications technology to bring her situation before the court of world opinion. Terri's parents posted videos of Terri "responding" to her mother, "laughing" at her father's jokes, and "watching" flashing lights and balloons to the Internet, where anyone and everyone weighed in with their opinions about her level of ability and consciousness. Terri's husband Michael maintained that such unauthorized sharing of his wife's image was an invasion of her privacy; in JG's case the Skype session between the hospital and JG's mother might also be seen in a similar light. Given their long period of estrangement, JG would likely not have chosen this particular moment or set of circumstances to re-engage with his mother.

JG did not leave an advance directive nor did he identify a health care surrogate. For persons in less traditional relationships and living arrangements, it is imperative that they have such documentation of their preferences for medical care, lest someone who no longer knows them well be given decision making authority (as happened in this case). Lesbians and homosexual partners who are not married or whose marriages are not recognized are cautioned to create such a paper trail so their life partners can make medical decisions (or even in some cases be allowed to visit) instead of a parent or other family member who may not approve of their union or be aware of the patient's wishes and desires. Creating such documentation requires a high level of health literacy (Araujo and Roscoe 2011), and given that he was uninsured and undocumented, it is not surprising that JG had not done so.

The laws in Florida specify a hierarchy of proxy decision makers for patients who are unable to make their own medical decisions, and the parent of such a patient is of a higher order than a cousin or an unmarried partner. JG was estranged from his mother in Mexico for many years, and even though she outranked the cousin in terms of Florida law, she was perhaps not the most suitable decision maker for her son. What was the hospital to do in this case? Pretend that JG's mother did not exist and therefore defy both state law and hospital policy? What they did do was aligned with the letter of the law, but not its spirit, which is to identify a person who either knows the patient's wishes and can use substituted judgment to make decisions, or knows the person well enough to decide what course of treatment might be in his or her best interest if they have not documented their treatment preferences.

Once a suitable proxy is identified, the appropriate process for decision making can be determined. One approach to end-of-life decision making is to weigh the relative benefit of a proposed medical treatment against whatever burdens accepting the treatment might entail. "Burden" need not be limited to the individual patient's experience, although this calculation must account for each patient's specific circumstances. Burden can include financial burden to the family—one is not required to accept treatment that would bankrupt one's family, for example. In JG's case, we have a skewed equation to solve; the benefit/burden ratio cannot apply in the same way when the family bears no financial responsibility for the patient's care. It would be heartless to withhold medical care from a person likely to recover some level of function and cognition because they lacked financial resources, but in JG's case, such an outcome is not possible. His mother's decision to continue aggressive life supportive care costs her nothing as she has no monetary assets to contribute, nor does she bear the burden of visiting her son or making any other arrangements for his care. She has made one decision that obligates the hospital to an indefinite financial burden, not to mention the burden of moral distress that the staff bears in caring for a patient whose care they believe is futile.

The decision put forward to JG's cousin and eventually to his estranged mother was whether ventilator support should be continued. For some families, discontinuing ventilator support is more straightforward than discontinuing tube feeding since ventilators are more "machine-like." Since JG was not able to be successfully weaned from the ventilator, he would have died quickly once it was removed, and no decisions about discontinuing tube feedings would have to be made. Human beings tend to associate food with comfort, and we wish to comfort and feed those we love, especially when they are ill. It is difficult to think about withholding food and water from our loved ones, even if doing so prolongs their lives in conditions they might find unacceptable. The idea that we might be "starving" our loved one is nightmarish, even though dying persons often choose to forgo eating and drinking as death approaches and the body begins shutting down. Medical science tells us that such a process is normal and comfortable, and that to force feed nutrients and liquids can increase the discomfort of a dying person by a considerable degree. A patient in a persistent vegetative state like JG is not able to perceive hunger, thirst, or any discomfort associated with either stopping or continuing tube feedings. That medical fact, however, did little to comfort Terri Schiavo's parents in the days after her husband successfully petitioned the court for permission to remove her feeding tube for the third and final time.

JG's case is one in which the spirit of our laws and ethical precepts governing and guiding end-of-life decisions come into conflict with the "letter of the law." The disconnects are many—a calculus of benefit and burden where one side benefits and the other is burdened; between a hospital's mission to provide compassionate high-quality care to all who enter and their obligations to be good stewards of their limited resources, such as money, space, and expertise. The disconnect between the Florida law's specification of biological relatives as proxy decision makers and those who might be in the best position to know the wishes of an incapacitated person, and the impracticality of relying on a document specifying medical care directives for persons who are undocumented and forced to live a shadowy life regarding their relationships with institutions such as hospitals and government agencies are also significant. There are disconnects in communication as well. Unlike many of the cases included in this book, here we seem to have helpful and clear communication between family members and hospital personnel, but still no straightforward resolution to this difficult situation. Hospital personnel, in following the letter of state law and institutional policy, were successful in identifying a family member higher up the proxy list than a cousin, alienating JG's cousin and eliminating his ability or willingness to either serve as a proxy or attempt to influence JG's mother's decisions. JG cannot speak for himself, but was nonetheless a participant in a dialogue with his mother that he may never have initiated. The use of technology was intended to bring closure, an attempt which backfired spectacularly, and started a new and difficult series of decisions with no straightforward resolution.

What is to happen to JG and others like him? This is far from an isolated case, as developed nations struggle to deal with a never-ending stream of immigrants seeking a better life away from strife, warfare, oppression, and limited economic opportunities. It is not likely that another lower cost health care facility can be found that would accept an unfunded patient with extensive needs for care and a long life expectancy such as JG, and it seems unlikely that JG's mother will change her mind about continuing his care in light of the fact that she bears no burden of inconvenience, financial obligation, or even the stress of attempting to have a continuing relationship with her son. It is likely that JG will develop pneumonia or a urinary tract infection, either of which could become life-threatening if antibiotic treatment is not initiated in a timely way. Is it possible that such a decision would become, in effect, a kind of "slow-code" in which antibiotics would not be immediately initiated? Again, it seems unlikely that JG's mother would decide that such medications should be withheld and under the present circumstances it seems she would need to be consulted. And what if JG dies, despite all aggressive treatments being brought to bear? His mother would likely lack the funds necessary to bring his body to Mexico, and the end to his story might be cremation at the city morgue and burial outside the city limits. It is a sad story, one made sadder by well-intentioned people attempting to follow well-intentioned policies, while the actual person in the bed goes unnoticed.

Notes

[1]See Dasta, Joseph F., Trent P. McLaughlin, Samir H. Mody, and Catherine T. Piech. 2005. Daily cost of an intensive care unit day: The contribution of mechanical ventilation. *Critical Care Medicine* 33: 1266–1271.

These authors' retrospective cohort analysis of the National Data Center Health's Hospital Patient Level database estimated the cost of intensive care for patients requiring mechanical ventilation at $10,794 for the first day, $4796 for day two, and $3968 for day 3 and beyond (in 2002 dollars). Those figures adjusted to 2016 dollars (using http://www.bls.gov/data/inflation_calculator.html) would put the cost of JC's first two weeks of care at close to $85,000, with an estimated cost of $37,000 per week subsequently.

[2]Jennett, Bryan, and Fred Plum. 1972. Persistent vegetative state after brain damage: A syndrome in search of a name. *The Lancet* 299: 734–737 (originally published as Volume 1, Issue 7753).

According to Jennett and Plum, patients with severe brain damage due to trauma or ischemia may survive indefinitely. Some never regain recognizable mental function, but recover from sleep-like coma in that they have periods of wakefulness when their eyes are open and move; their responsiveness is limited to primitive postural and reflex movements of the limbs, and they never speak. Such patients are best described as in a persistent vegetative state, which should be clearly distinguished from other conditions associated with prolonged unresponsiveness. What is common to these patients is the absence of function in the cerebral cortex as judged behaviorally.

[3]The upper airway warms, cleans and moistens the air we breathe. The trach tube bypasses these mechanisms, so that the air moving through the tube is cooler, dryer and not as clean. In response to these changes, the body produces more mucus. Suctioning clears mucus from the tracheostomy tube and is essential for proper breathing and to prevent a chest infection if the secretions left in the tube become contaminated. Therefore, there is a logical inconsistency in JG's mother's reasoning since continuing ventilator support would necessarily entail continued suctioning.

[4]The Multi-Society Task Force on PVS, (1994). Medical aspects of the persistent vegetative state. *The New England Journal of Medicine, 330,* 1572–1579. The Task Force reported the mortality rate for adults in a persistent vegetative state after an acute brain injury such as JG's as 82% at three years and 95% at five years; approximately 90% of patients died within 10 years.

[5]Migration Policy Institute (MPI) Data Hub, State Immigration Data Profiles Chart, **Florida—Demographics and Social,** *sources*: Migration Policy Institute tabulations of data from the U.S. Census Bureau's American Community Survey (ACS) and Decennial Census. Unless stated otherwise, 2014 data are from the one-year ACS file. Estimates from 1990 and 2000 Decennial Census data as well as ACS microdata are from Ruggles, Stephen, Matthew Sobek, Trent Alexander, Catherine Fitch, Ronald Goeken, Patricia Hall, Miriam King, and Chad Ronnander. 2011. Integrated public use microdata series: Version 4.0. (Machine-readable database, Minnesota Population Center [producer and distributor]; downloaded from www.migrationpolicy.org on 10/28/2016).

[6]Migration Policy Institute Data Hub, Unauthorized Immigrant Population Chart, **Profile of the Unauthorized Population: Florida,** *source*: Migration Policy Institute (MPI) analysis of U.S. Census Bureau data from the 2014 American Community Survey (ACS), 2010–2014 ACS pooled and the 2008 Survey and Income Program Participation (SIPP) by Colin Hammar and James Bachmeier of Temple University and Jennifer Van Hook of the Pennsylvania State University, Population Research Institute; downloaded from www.migrationpolicy.org on 10/28/2016.

References

Araujo, Meagan, and Lori A. Roscoe. 2011. Meaning in context: The real work of medical interpreters. In *Contemporary Case Studies in Health Communication: Theoretical & Applied Approaches*, ed. M. Brann, 47–58. Kendall Hunt: Dubuque, IA.

Molina, Natalia. 2011. Borders, laborers, and racialized medicalization: Mexican immigration and U.S. public health practices in the 20th century. *American Journal of Public Health* 101: 1024–1031.

Roscoe, Lori A., Hana Osman, and William E. Haley. 2006. Implications of the Schiavo case for understanding family caregiving issues at the end-of-life. *Death Studies* 30: 149–161.

Case 14—To Treat…or Not to Treat?

CB was a 45-year-old man who was admitted to the head and neck cancer service of a large, tertiary care hospital immediately upon presentation to the emergency department (ED) where he complained of severe head pain. He reported a 9-month history of pain on the left side of his head behind his left eye, with slight diplopia (double vision) and left otalgia (ear pain). He had a long history of tobacco and alcohol use, and it was evident that he had been drinking when he arrived at the hospital. The physical exam revealed a large mass in the left neck underneath the sternocleidomastoid muscle from the angle of the mandible to the level of the clavicle; a CT scan revealed this mass to be approximately 5.3 cm × 6.4 cm in size and to extend from the clavicle inferiorly to the skull base superiorly. The carotid sheath was completely surrounded by the tumor, which also abutted the left optic nerve; the scan also indicated the tumor was beginning to erode the skull base.[1] The clinical exam further revealed bilateral neck adenopathy[2]; a fine needle aspiration (FNA) of one of the large nodes came back positive for squamous cell carcinoma.[3]

CB's case was discussed in the weekly interdisciplinary tumor board conference.[4] Based upon the clinical findings, the scans and the lab report of the FNA, the attending head and neck surgeon diagnosed this as Stage IV disease and judged the tumor to be inoperable because of its size, location and extensive nature.[5] She also felt that chemotherapy would be ineffective in this particular case and estimated CB's likelihood of survival at five years to be less than 10%. She stated her opinion that the patient would survive perhaps no more than two months without treatment, but that surgery to "debulk" the primary tumor and neck nodes could possibly extend his life by three to four more months. The radiation oncologists pointed out that radiation therapy could be expected to shrink the size of the mass and nodes, thus also extending his life a few extra months while adding the benefit of pain relief from tumor shrinkage; the surgeons and medical oncologists concurred with this view. All physicians present agreed, however, that CB would eventually succumb to his disease.

Nevertheless, the medical team was divided on the appropriate course of action in this case. Even in situations where patients are seen by physicians to have little

© Springer International Publishing AG 2017

L. A. Roscoe and D. P. Schenck, *Communication and Bioethics at the End of Life*,
https://doi.org/10.1007/978-3-319-70920-8_14

chance of survival, it is customary to offer them either (a) a choice of indicated treatments, or (b) no treatment for the disease itself, but at least palliative care to address symptoms and improve quality of life for the patient. There were those in the tumor board meeting who felt strongly that CB should be offered precisely that, which is to say that he should have carefully laid out for him the pros and cons of the three options discussed in tumor board: (1) debulking surgery, (2) radiation therapy (XRT), or (3) no treatment except for palliative pain and comfort measures only. It would then be left up to him, in further consultation with the attending physician, to make an informed choice. However, there were complicating factors.

CB's only home was a room above a bar loaned to him by the owner in exchange for maintenance work. Given that fact, along with his history of tobacco and alcohol use over many years, there were those who felt he was not likely to follow his complicated pre-operative care instructions as well as the post-operative self-care that would be required were surgery to be at all successful. Team members espousing this view also expressed concern that the proposed surgery could last as long as twelve hours, that this would mean dedicating an entire day of one operating room to one case only, and that such valuable OR time could otherwise be shared among several other needy patients who might be more likely to do well post-operatively than CB. These team members also pointed out that were he to be offered radiation therapy instead of surgery, the advanced nature of his disease would probably require placing him ahead of other patients in the queue, and that those waiting their turn in line would now have their treatments delayed. Consequently, these team members believed it inappropriate to offer CB all three options but thought that the palliative care option would be best.

The attending head and neck surgeon happened to be one of those in this second group. She argued strongly for her position, but she also listened carefully to the views of the other group, the group that felt CB should be offered all three options, with the choice then left to him. In the end, she indicated she would review all that had been discussed before deciding what to present to her patient.

Discussion Questions

1. Before reading either of the two following responses, formulate your own position on the option(s) that should be offered to CB.
2. *Autonomy* is often viewed as the overriding principle in bioethics, which is to say that it is viewed as the equivalent of saying that if one is medically competent one has the right to decide freely for oneself. Do you agree that this should always be the case? Could there be valid exceptions?

A Bioethicist Responds

It is not difficult to imagine the give and take that took place during the discussion of this case in tumor board:

"You just want to deny him the difficult, time-consuming, costly surgery because he's a charity case, he's an alcoholic and he's never taken care of himself! Well, he's a human being and worthy of our care just like anyone else!"

"Look, I've spent countless hours in the OR on poor, indigent folks who had no financial means and who may have done it to themselves (i.e., caused or aggravated their disease through excessive alcohol and tobacco use), and I'd do it again, but I believed they had a chance because I felt they'd be able to follow through after surgery with what they'd need to do to help heal themselves. But in this case I have no confidence he'll do that, and so I'm really afraid all the work we'll do will be for naught!"

"Yeah, but you don't know that, so we have to give him the same chance as anybody should have. Besides, if you don't offer him all the options, you're not holding true to the rules of informed consent!"

"Well, what's more important, sticking to some formal rule, or doing what is best for the patient in the long term? This patient has no way of knowing or understanding what will happen to him, what he will truly experience, what kind of suffering he will go through if he should choose the wrong option, and I do know. I don't want to be paternalistic, play God, or say I know what's good for him, but in this case I do know what is best for him because of my experience with patients who lack sufficient social support."

The primary ethical issues that quickly surface here are patient autonomy and justice. Autonomy in this case includes questions related to informed consent and paternalism; justice here must account for both fairness and proper use of scarce resources.

We turn first to a quick review of the essential elements of informed consent, which are: 1—the patient must possess the medical *competence* to understand and follow the standard consent process; 2—the patient acts completely *voluntarily*; 3—the consent form and process must include complete *disclosure* of all aspects related to the procedure (surgery, etc.) to which the patient is consenting; 4—the physician must make a *recommendation* as to what he/she honestly believes is the best treatment option (including no treatment) for this specific patient; 5—the patient must ideally have an *understanding* of what has been disclosed and recommended; 6—the patient makes a free and unfettered *decision*; and finally, 7—the patient, or patient's legally authorized representative, formalizes the decision by means of an *authorization* (written or oral).[6]

In view of the preceding it is easy to see that were the attending physician not to consent CB properly, he could not possibly make a truly *informed* decision about any option she might propose to him. Were she not to let him know that there existed three typical therapeutic options for dealing with his particular disease, carefully explain the pros and cons of each, tell him what she thinks is his best option based on her knowledge and experience, and then let him choose which seemed right for him after reflective discussion with her, he could not possibly be fully and fairly informed before making his decision. In fact, if she were to do as she had first suggested and simply offer him palliative care, she would be acting paternalistically.

Most persons today are strongly averse to paternalism practiced by either men or women unless they believe that a specific situation is so unique that the intentional overriding of a person's autonomy may be justified by the goal of beneficence they hope to attain in taking that action. This is risky, however, and there must be very compelling reasons for doing so. Perhaps one such compelling reason in this case could be the very fact of limited resources, combined with the attending physician's conviction that her long hours of work in the OR would likely go for naught when her patient found himself unable to care for himself post-operatively, with no adequate support system to care for him either, including the likelihood of an early death anyway. The challenge then would be to demonstrate that a fair allocation of resources would permit overriding this patient's autonomy to the extent suggested here where strong paternalism comes into play and any informed consent process that may ensue virtually takes on a whole new meaning.

The whole question of consent remains an interesting if not challenging one. Debate continues as to whether truly informed consent is ever really obtained from a patient. Then there is the question of whether it really matters if true informed consent is obtained, or whether what really is at stake is doing the best one can and making sure that a record of the process has been made. Moreover, some studies have shown that what may really matter in the final analysis, at least to patients, is not the more formal or legalistic elements physicians and bioethicists generally think of with regard to informed consent, but rather such things as enhancement of trust through the referral process, the idealization of the doctor, the belief in the doctor's expertise rather than the medical information he or she might be able to provide, and that when a patient accepts the doctor's recommendation for a treatment he or she is, in effect, thereby giving his or her consent (McKneally et al. 2009). Yet while this may offer some perspective, it may not provide much practical guidance ethically.

Full disclosure is in order at this point. The present writer was a member of the tumor board in question and was one of those who advocated, along with the attending head and neck surgeon, in favor of recommending only palliative care to CB. It genuinely appeared that this would be in the best interest of this patient; that the co-morbidities associated with a long surgery would do far more harm than good; that he could not possibly care for himself afterward; that there was little likelihood of anyone else providing care for him post-operatively; that there would be a good chance of his being lost to follow-up; that if he went for XRT and jumped ahead of others in the queue it would be grossly unfair to them; that the amount of extra time afforded him through surgery or XRT would be so little (several months at best) as to be relatively negligible since there would have to be trade-offs for discomfort; there might be the possibility of getting him into hospice care, with nurses and other allied health professionals who could be assured of visiting him at his current residence or in a skilled nursing or rehabilitation facility if it came to that. He was surely going to die within the foreseeable future. It seemed that helping him to do that as quickly and comfortably as humanly possible was the most beneficent and ethical choice; this option would also ensure the best use of the scarce human and material resources available to the institution while concomitantly observing issues of fairness and beneficence regarding other patients for whom the institution was responsible.

Without further consultation with anyone, however, the attending physician decided to play it "by the book." She presented CB with the standard three options; explained to him that he was not a realistic candidate for chemotherapy and why; laid out the seriousness and challenges of a very long, possibly twelve-hour operation; carefully presented the pros and cons of the surgery; told him what he would be required to do for himself on a daily basis, and for how long, after the operation; explained that he would definitely also need a caregiver for some weeks right after surgery; explained all that would be involved with XRT and what he could expect with that; told him approximately how many additional months of life he might expect with either surgery or XRT, but that these additional months would come at a cost of additional discomfort; told him what was meant by hospice and palliative care, approximately how many months of life he might expect if he chose that route, and told him this would be the most pain-free choice and why. She stressed that no matter which choice he made, she and the rest of the medical team would be with him until the end and that they would ensure that he would be kept as free of pain and as comfortable as possible. Lastly, she told him what she would choose if she were in his place and why—that enrolling in hospice and taking a palliative care approach would most be most likely to maintain his comfort, reduce suffering, and enhance his quality of life. She reported several days later that after listening carefully to all that she had said, CB merely smiled gently at her and said, "Thanks, doc. Glad to hear you agree with me. Truth is, I'd pretty much made up my mind that's what I wanted anyway. I just wanted to hear you say it first!"

A Health Communication Scholar Responds

Communication about end-of-life care is increasingly recognized as a core clinical skill, but some doctors, unlike the attending physician in this case, are not well prepared to have these conversations. This case is an example of the ways in which honest, personal, face-to-face communication between a seriously ill person and a physician can and should occur. The attending physician in this case likely had strong feelings about the correct medical course of action, and equally strong feelings about the correct ethical course of action. There was a chance, of course, that doing the right thing ethically might complicate doing the right thing medically. Physicians often find themselves in situations where they have presented the options available to the patient, some of which have very limited likelihood to improve the patient's life expectancy, functional status, or quality of life, only to have the patient and his or her family declare that "everything" must be done! Even though that was not the case with CB, allow me to speculate on what the attending physician in this case likely feared was at stake when she made the choice she did to fully inform CB of his choices for treatment.

Crisis situations at the end of life, such as when a patient with a poor prognosis requests that "everything" be done, pose serious communicative challenges (Roscoe and Tullis 2015). Surely this was one scenario that crossed the mind of the attending physician in CB's case. Decisions on the part of a patient or family to do "everything" are often seen as a demand for care that may be burdensome or even harmful, rather than the start of an important conversation about values and goals. We can surmise from the skillful communication demonstrated in this case, that rather than defer to a demand that "everything" be done, the attending physician would have continued the conversation about what "everything" meant to CB. This is how it should be. Informed consent, as presented in the *Bioethicist's Response*, requires information, but it also requires an engaged, conversational exchange to allow each individual's motivations, concerns, and convictions to emerge and be addressed.

Situations where so-called "bad news"[7] must be conveyed to patients have prompted the development of prescriptive approaches to communicating. Such scripted approaches are better than ignoring the need for such a conversation, but they have limitations since they focus more on the physician's message ("this is how sick you are and these are the decisions you need to make") than on developing a relationship with the patient so as to elicit the meaning of the illness and its treatment, identify the patient's values and goals, and jointly agree on a way forward. Whereas older models of the communication process focused on message transmission, more current models of communication are less about delivering messages and more about the ways in which meaning is co-constructed in relationships. The words a physician uses, his or her body language and tone of voice, the time and care taken—all contribute to and shape the illness experience and context for decision making for the patient. In this case we have an example of a physician who entered the world of her patient, and explained in detail how CB's limited life expectancy would likely play out under various treatment scenarios. This is difficult and brave work. There are such obvious chasms to transcend—the physician is healthy, the patient is not; the physician is well- educated, the patient is not; the physician is well-off financially, the patient lives in a room above a bar; the physician has detailed knowledge of treatments and their outcomes, the patient has limited health literacy; and the list could go on. Despite these obstacles, the attending physician in this case did bridge those gaps, and managed to develop a relationship with a very sick man who had a difficult, high-stakes choice to make.

I wish to encourage every physician reading this case to accept the inherent risks inconveniences and spend the time necessary to forge a relationship with his or her patients when important treatment decisions must be made. Unfortunately, not all physicians (or any of us, really) have the communication skills that allow us to do so. For physicians, communication training most often takes the form of tools and scripts, which do not fit the nuances inherent in any difficult patient situation. I want to encourage physicians to do what the attending physician did in this case—engage the patient in an honest dialogue, and do not stop until a real "meeting of the minds" has taken place. It involves more than making eye contact, or following a series of steps to effectively "deliver the bad news." All of us need human relationships in order to thrive (and to die with dignity), and seriously ill people desperately need such a connection to and conversation with their physicians in order to make the decisions they need to make. Especially in the case of CB, who is alone, the doctor's role in providing a communicative space is all-important.

All patients deserve such engaged care, but head and neck cancer patients face specific communicative challenges. Patients with head and neck cancer, whose disease processes and surgical treatments are often disfiguring and limit their ability to communicate, may be at higher risk of having insufficient information about end-of-life care (Roscoe et al. 2013; Schenck 2002). Since their oncologists (including surgeons, and medical and radiation oncologists) and other specialized clinical health professionals (nurses, psychologists, and social workers) are often the only ones who really understand the nature of this disease, its potentially disfiguring treatments, the outcomes of various treatment modalities, and often their patients' disadvantaged access to health care resources over their lifetimes, they have a heightened responsibility to ensure that their patients understand their prognoses and have the information necessary to make informed decisions about end-of-life care.

The question remains whether such "empathetic witnessing" as described by Broyard (1993) and others and the compassionate communication accompanying it can be taught to physicians and other medical practitioners. Perhaps medical students should be admitted based on their emotional intelligence along with their MCAT scores. More pedagogical research can be conducted to try to discern when and in what ways the medical school curriculum can best accommodate communication skills training and practice. It has been documented that empathy declines over the course of one's medical education, but are there ways that can be changed? These are open questions, subject to much debate among medical school faculties and curriculum committees.[8]

Most of us feel that we are good communicators, but as this case demonstrates, what really matters is that our conversational partner is satisfied with our exchange. CB's comment about "wanting to hear it from you first" is a good indication that this communicative exchange was beneficial to his decision making and peace of mind. We can be fairly sure that he grasped not only the information about his disease and treatment options, but also the warmth and empathy of the attending physician. Most of us are quite cavalier about this important aspect of good communication—we rarely check for grasping. If we were passing a physical object—picture runners in a relay race passing a baton—we would not let go until we felt the reassuring pressure that indicated that our receiver had a firm grasp. We do this with important information all the time, and physicians are no less culpable: We put the information in the vicinity of the other person, and hope they get it. What is clear is that when such good communication does occur, the dignity and personhood of the patient is affirmed, the moral distress of physicians and other health care providers is lessened, and scarce resources are justly managed.

Notes

[1]The anatomical area of the head outlined in these two sentences corresponds roughly to what is known as the *infratemporal fossa* (for detailed drawings, see: Netter, Frank H. 2014. *Atlas of Human Anatomy*, 6th ed. Philadelphia., PA: Saunders/Elsevier). Two things should be noted at this point. The first is that the tumor appears to be eroding the skull base, which means that the brain may soon be invaded by cancer. The second is that the indication of tumor completely surrounding the internal carotid artery sheath suggests it is only a matter of time before the cancer weakens the wall of the artery itself sufficiently to cause what is known as a "carotid blow-out." This would mean virtually instant death, or death within relatively few seconds. The patient could be rendered unconscious almost immediately, but the event would doubtless be horrifying for others alongside, especially loved-ones, for the weakened vessel is often near the surface of the skin when this occurs, causing blood from the ruptured artery to continue spurting until sufficient blood volume is lost and/or the heart begins to slow.

[2]*Adenopathy* indicates the presence of enlarged lymph nodes or nodal disease. Nodes are usually first noticed on physical exam in clinic, but only if they are large enough to be palpated. They are more easily seen on CT, MRI or PET scans. Not all enlarged nodes are necessarily malignant, however. There are a great many lymph nodes in the head and neck, located along "chains," and their main purpose is to serve as filters to collect waste products, invading germs or other foreign "poisons" in the body. They will also pick up errant cancer cells from adjacent tumors, or cancer cells released from distant sites in the body.

[3]This chapter contains greater medical detail and terminology than most of the other parts of this book, which in itself reflects the complex nature of the surgical subspecialty known as *otolaryngology-head and neck surgery*. The anatomy of the head and neck is highly complex; cancer terminology can be confusing; cancer staging of the head and neck is complicated; and, the field of cancer therapeutics is a lexicographer's dream.

[4]These conferences ("tumor boards," for short) generally consist of a team of head and neck surgeons, radiation oncologists, medical oncologists, a radiologist, a pathologist, residents and/or fellows, the nurses who work with each of the clinical medical specialists, speech-language pathologists, nutritionists and social workers involved in the care of head and neck cancer patients. Depending upon the institution, the tumor board may also be attended by students (e.g., medical, nursing), pastoral care personnel and a bioethicist. Everyone present is bound by HIPAA regulations. The purpose of the tumor board is to review cases, to seek input primarily from physician members of the team and to develop a treatment plan. In all cases, however, the final decision on a treatment plan is the responsibility of the attending together with her/his patient.

[5]For staging in head and neck cancer, see: Deschler, Daniel G., Michael G. Moore, and Richard V. Smith (eds). 2014. *Quick reference guide to TNM staging of head*

and neck cancer and neck dissection classification, 4th ed. Alexandria, VA: American Academy of Otolaryngology-Head and Neck Surgery Foundation.

[6]The best overall treatment of informed consent is still to be found in Faden, Ruth R., and Tom L. Beauchamp. 1986. *A history and theory of informed consent*. New York: Oxford University Press. See also Beauchamp, Tom L. and James F. Childress. 2013. *Principles of biomedical ethics*. 7th ed. New York: Oxford University Press.

[7]For more information about scripted communication models for giving "bad news" to patients, see:
Arnold Back, R. M. Arnold, W. F. Baile, James A. Tulsky, and K. Fryer-Edwards. 2005. Approaching difficult communication tasks in oncology. *CA: A Cancer Journal for Clinicians 55*: 164–177; and Eggly, S. S., Terrence L. Albrecht, K. Kelly, H. G. Prigerson, L. K. Sheldon, and J. Studts. 2009. The role of the clinician in cancer communication. *Journal of Health Communication, 14*: 66–75.

[8]For more information on preserving empathy in medical students and physicians in training, see:
M. Hoja, M. J. Vergare, K. Maxwell, et al. 2009. The devil is in the third year: A longitudinal study of erosion of empathy in medical school. *Academic Medicine, 84*:1182–1191.
B. W. Newton, I. Barber, J. Clardy, E. Cleveland, and P. O'Sullivan. 2008. Is there a hardening of the heart during medical school? *Academic Medicine, 83*:244–249.
S. Rosentha, B. Howard, Y. R. Schlussel. D. Herrigel, G. Smolarz, B. Gable, J. Vasques, H. Grigo, and M. Kaufman M. 2011. Humanism at heart: Preserving empathy in third-year medical students. *Academic Medicine, 86*:350–358.

References

Broyard, Anatole. 1993. Intoxicated by my illness and other writings on life and death. New York: Ballantine Books.

McKneally, Martin F., Douglas K. Martin, Esther Ignagni, and Jason D'Cruz. 2009. Responding to trust: Surgeons' perspective on informed consent. *World of Surgery* 33: 1341–1347.

Roscoe, Lori A., and Jillian A. Tullis. 2015. The meaning of everything: Communication at the end of life. *Journal of Medicine & the Person* 13: 75–81.

Roscoe, Lori A., Jillian A. Tullis, Richard R. Reich, and Judith C. McCaffrey. 2013. Beyond good intentions and patient perceptions: Competing definitions of effective communication in head and neck cancer care at the end of life. *Health Communication* 28: 183–192.

Schenck, David P. 2002. Ethical considerations in the treatment of head and neck cancer. *Cancer Control* 9: 410–419.

DJ was a 55-year-old man who arrived at the emergency department of a community hospital with a high fever and altered mental status. Doctors admitted him for treatment of septic shock. He received intravenous fluids and antibiotics, as well as vasopressors to help maintain adequate blood pressure. Blood culture tests revealed polymicrobial pathogens, a combination of viruses, bacteria, fungi, and parasites. A transesophageal echocardiogram revealed vegetation on the patient's aortic valve, an indication of infective endocarditis. DJ was transferred to a tertiary care hospital to be evaluated for cardiothoracic surgery and valve replacement.

DJ's medical history revealed that a year earlier he had some dental work done "in a garage," after which he developed endocarditis and had an aortic valve replacement. Two years prior to that surgery, he had been seen in another local hospital with left lower leg cellulitis and endocarditis, for which he was prescribed a 6-week course of antibiotics. DJ discontinued taking the prescribed medication after three weeks because he started to feel better. DJ also had a history of chronic pain syndrome and fibromyalgia, which had been treated over a period of 13 years with Duragesic patches and Dilaudid dispensed from a pain management clinic, until new state legislation put the clinic out of business. DJ had also been diagnosed with bipolar disorder for which he was taking no medication. He had other surgeries on his right knee and ankle, his left wrist and hip, as well as a splenectomy (removal of one's spleen, an organ that helps fight infection and filters damaged blood cells), which was required after he had sustained a gunshot wound.

The day after DJ's admission to the hospital, he was seen by a cardiothoracic surgeon, who found a used syringe under DJ's bed. DJ admitted to using a variety of narcotics, mostly injecting heroin, in attempts to better control his pain and mood swings. The surgeon documented in DJ's chart that he believed redoing the valve replacement when DJ continued to inject heroin (thus greatly increasing the risk of reinfection and need for repeated heart valve replacements) was futile. The surgeon recommended that DJ be given a prescription for antibiotics and be discharged since he was not a good candidate for surgical intervention. The attending physician also requested a consult with an infectious disease specialist, who documented that

© Springer International Publishing AG 2017

L. A. Roscoe and D. P. Schenck, *Communication and Bioethics at the End of Life*, https://doi.org/10.1007/978-3-319-70920-8_15

the polymicrobes found in DJ's blood cultures were extremely resistant to antibiotics. The infectious disease specialist recommended a 6-week course of intravenous antibiotics, which were to be started immediately and which could be administered in a skilled nursing facility instead of the hospital.

The social worker assigned to DJ's case attempted to find an available Medicaid bed for him in a nursing home as DJ did not have health insurance. Meanwhile, he developed fungemia (the presence of fungi or yeasts in the blood) and was started on intravenous antifungal medications in addition to the intravenous antibiotics. During this time, DJ was only intermittently responsive and he appeared even more lethargic than would have been expected given the relatively modest doses of pain medication he had been receiving. More used syringes were discovered in DJ's room, along with a knife, two spoons, three empty bottles of Dilaudid, and two nearly empty bottles of vodka. "It's my girlfriend," DJ told the social worker, "she's the one who brought in the drugs and alcohol! I didn't touch anything!".

An ethics consult was called the following week by the attending physician, which was also attended by the infectious disease specialist and the cardiothoracic surgeon. The infectious disease specialist advocated strongly in favor of surgery as the best treatment for DJ, recommending that it be scheduled after three negative blood culture test results indicated an absence of infection.[1]

The surgeon again expressed reservations, stating not only his concern for the high risk of mortality and complications, but also the high risk of reinfection, were the patient to continue injecting drugs. He bolstered his position by saying that most other cardiothoracic surgeons would refuse to perform surgery on this patient under the present circumstances. He finally agreed that if the antibiotics and antifungal medications were effective, and if three consecutive negative blood cultures were obtained, he would perform the surgery, but only if DJ signed an agreement stating that he would not continue to inject heroin. The ethics committee endorsed the proposal of asking DJ to sign an agreement, stating they found no medical contraindications to the surgery.

DJ remained in the hospital and received the intravenous antibiotics and antifungal medications. His room was monitored carefully, and no additional used syringes or liquor bottles were found. His mental status and lethargy improved somewhat, and three negative blood cultures were obtained after the full course of intravenous antibiotic treatment. As part of the consent process for the valve replacement surgery, DJ signed an agreement stating he would no longer inject drugs. The cardiothoracic surgeon was consulted again, and although he reinforced his earlier position that he believed DJ would be unlikely to stop injecting drugs despite the agreement, he agreed to perform the surgery.

After a complicated post-operative course, DJ was discharged to a rehabilitation facility, where he received physical and occupational therapy. He also met at least once with a substance abuse counselor on retainer at that facility, and he denied using any injectable drugs, including heroin. DJ was discharged home after 6 weeks.

Shortly thereafter, his girlfriend brought him back to the hospital's emergency department after finding him unresponsive. The work-up revealed complications from a methicillin-resistant *Staphylococcus aureus* (MRSA) endocarditis involving his new aortic valve.[1] Despite aggressive intravenous antibiotic therapy, DJ's confusion worsened and he was diagnosed with multi-system organ failure and severe sepsis. He died later that evening.

Discussion Questions

1. Imagine that you are a member of the ethics committee. What would you recommend in this case?
2. How can physicians provide care to patients such as these, when they present with legal and social problems in addition to serious medical problems?
3. Is it ethical to require patients to sign agreements about their post-operative care?
4. When making medical decisions, are there differences between allocating scarce resources such as organs versus surgeons' time and expertise?
5. Do patients have a "right" to medical treatment? Is access to health care a privilege or a right?
6. Under what conditions might our access be rightfully limited?
7. Are there situations in which physicians should be required to provide futile care?

A Bioethicist Responds

DJ's story contains elements regrettably found in numerous patient stories today. It is the story of a patient who has become dependent upon, if not addicted to, prescription pain medication; who was unfunded; who suffered from chronic health problems; who had undergone several, doubtless expensive, medical interventions; who had apparently failed to deal with his bipolar disorder, either because of neglect or because of inability/unwillingness to afford or tolerate appropriate medication(s) for it; and, who nevertheless continued his drug habit after hospital admission. Patients like DJ are not popular in hospitals, and few physicians are eager to have them in their charge. It may therefore not be surprising that the cardiothoracic surgeon went to the lengths he did before agreeing to perform the heart valve replacement, an agreement he made conditional upon the following: (1) DJ's completion of a course of antifungal/antibiotic treatments; (2) three consecutive negative blood cultures following this course of treatments; and, (3) DJ's signature on an agreement that he would no longer inject drugs. The sad outcome of this case aside, one would have relatively little difficulty constructing an argument in support of the surgeon's position were the defense of it to be established largely on the medical facts of the case, DJ's medical and behavioral history, and the principles of justice, nonmaleficence and beneficence (which together could be shown to outweigh DJ's demand for autonomy), especially if justice were emphasized.[2]

Approaching the case in such a manner would lend validation to the surgeon's position, particularly if efforts were made to eliminate any personal bias of DJ himself while assessing strictly the facts, history and principles mentioned above. Nonetheless, such an approach would still be unsatisfying, for at the very least nothing would have been said of the virtues, those habits of being that truly give the medical enterprise its human character. The virtues must certainly be taken into account, but an account of the virtues alone in this case might still not be acceptable to those used to a broader perspective. A narrative approach might be very useful, but it would not be sufficiently normative to this writer's mind. Should we opt for a case-based approach? Possibly, but selecting a particular "method" of analyzing this case is perhaps to pose the wrong question in the first place. I suggest that the more cogent issue has rather to do with the very question of *rights* themselves, the rights of both patient and physician, which is to say, of both DJ and the cardiothoracic surgeon.

The debates over *patient rights to refuse treatment, patient rights to treatment,* and *physician rights to refuse to treat* (a patient or disease) are legion. Those related to patients refusing treatment have generally focused on issues of autonomy, although questions of informed consent, which therefore raise questions of beneficence and nonmaleficence by implication, have also been common. This is such a well-known subject area that it needs no further discussion here.

The subject of *patient rights to treatment*, strictly speaking, has received far less explicit attention over the years. In fact, it is difficult to identify much in the literature that speaks directly to a patient's "right" to treatment. In 2014, Lepping and Raveesh published an interesting piece that dealt with involuntary detention and treatment in psychiatric care, an issue that had been, as they noted, "subject to endless legislation, campaigns, criticism and ethical debate across the globe" (Lepping and Raveesh 2014, 1). Reference to an article focusing on psychiatry and involuntary detention of patients in the context of DJ's case may seem odd, but I mention it here because I believe it will serve, albeit ironically, as a kind of "inverse" support for a point I will emphasize later in the context of care, and because it leads directly into the thrust of Lepping's and Raveesh's argument, which is: "A psychiatric practice less concerned with the primacy of autonomy would more seriously consider the patient's relationships, their care needs and their long-term social contexts. It would give more importance to the opinions of significant persons in the patient's life, and consider these opinions to form a view of the patient's best interests that is not merely based on theoretical wishes and aims" (Lepping and Raveesh 2014, 2). While Lepping and Raveesh would in no way run rough shod over the autonomy of a patient, even one severely mentally ill, they would insist on taking careful account of the wider network of relationships in which the patient lives and to which he or she must ultimately return post-treatment. In other words, the patient's relationship(s) to another, not to mention to her/his broader environment, is of singular importance.

Meyerson (2015) directly addressed the question of *patient rights to treatment* and examined the claim that patients have a "right" to an innovative surgical procedure that is yet to receive full FDA approval.[3] Following considerations of patients in both terminal and non-terminal conditions, positive and negative rights, a proportionality analysis (such as the point at which one person's rights usurp the rights of another), and competing interests of various kinds, Meyerson calls for a middle road between the conservative view, where an oversight committee could automatically deny a procedure request, if it believed an innovative procedure to be too risky (despite the patient's assessment that the risks were worth taking), and the liberal view, where a patient's wishes would always be respected. Meyerson argues her point on the basis of proportionality, noting that the choice to undergo innovative surgery is an intensely personal one that is made in an effort to preserve one's life or to protect one's health, and that the arguments for blocking access must, therefore, be sufficient to demonstrate that what is to be gained by prohibition is sufficiently important to compensate for interference with patient rights (Meyerson 2015, 350).

The subject of *physician rights to refuse to treat* could be traced into ancient and medieval history, an unnecessary exercise for present purposes, but it is worth recalling the discussions that developed as a result of widespread HIV infection. Norman Daniels emphasized some of the most important points on this specific topic regarding nosocomial risks, justice, professional obligations of physicians to patients, the concept of the virtuous physician, and the limits of a duty to treat (Daniels 1991). The central issue for Daniels, after due consideration of each of the foregoing points, was that a just society requires adequate access to treatment be available for HIV patients. He acknowledged this will mean our having to overcome certain financial, personnel and geographical barriers, which may lie at the source of physician reluctance to treat HIV patients, but he concluded by stating that, "we can best reaffirm that physicians have a duty to treat despite nosocomial risks if we show our social commitment to assure access to care in all feasible ways" (Daniels 1991, 46). Daniels' article was born of a particular set of historical circumstances, but his emphasis on the physician's duty to treat (albeit through our social commitment to assure access to care) is noteworthy.

Nonetheless, positions virtually to the contrary are espoused by many others, especially where it is felt that physicians are not ethically obligated to provide care considered to be futile, unreasonable, or simply not medically indicated (Luce 1995).

And yet, there are those who seriously challenge any suggestion that physicians should refuse to treat patients—physicians' best medical judgments, nonmaleficence, futility, concerns over resource allocation, or physician autonomy notwithstanding. Philosophical arguments have been offered to defend the position that it is competent patients who should decide whether or not they wish to have certain treatments, particularly when physicians raise futility arguments, and despite the likelihood of a treatment's success or of a patient's chances of survival (Wreen 2002). Others have argued more pointedly that the basic model for decision making should put the patient and his or her values at center stage, and that "in futility cases, short of certainty that a result desired by the patient cannot be achieved, the question of futility is not one for which HCPs (health care professionals) should be the ultimate judge" (Gampel 2006).

In my view, none of these approaches offers much help with respect to the case of DJ, for none takes sufficiently into account the relationship between physician and patient that is absolutely central to the fact of illness and the practice of medicine, not to mention the concept of "beneficence-in-trust" that must ground the physician-patient relationship (Pellegrino and Thomasma 1988). Readers will also find a very useful, albeit short, treatment of this concept in Sulmasy's *Forward* to a 2001 collection of some of Pellegrino's seminal writings (Bulger and McGovern 2001). Central to beneficence-in-trust, and to a possible understanding of DJ's situation, is what Pellegrino and Thomasma mean by the term itself: "By beneficence-in-trust we mean that physicians and patients hold "in trust" (Latin, *fiducia*) the goal of acting in the best interests of one another in the relationship" (Pellegrino and Thomasma 1988, 54). What this ultimately means is that the physician and the patient have responsibilities not only toward one another, but that the patient also has responsibilities for her/his own health; it is not the physician alone who is responsible for the best (health) interests of the patient. Pellegrino and Thomasma do indeed make clear that beneficence-in-trust is their attempt to describe a middle road between paternalism and autonomy, but they also make it clear that, "since one cannot heal by neglecting or overriding patient wishes, physicians must be stewards of patient values and preferences whenever possible" (Pellegrino and Thomasma 1988, 202).

The problem for DJ's physicians, however, and most especially for the surgeon asked to perform the valve replacement, is that there is no evidence that any of them developed a relationship with DJ sufficient to where they could possibly have become adequate stewards of his values and preferences. What little any of the health care professionals in the hospital knew of DJ was that he had proven himself to be unreliable; moreover, the surgeon had stated early on that he felt doing the valve replacement would be futile as long as DJ continued to inject drugs. I doubt even Pellegrino would have disagreed with the surgeon on this point, and yet the surgeon appears to have given into pressure from the infectious disease specialist. That action alone is cause for ethical concern, as it calls into question either the strength of the surgeon's own convictions or the motivations for his first refusal. But, of even greater ethical concern is his insistence on the agreement DJ had to sign as a condition for the valve replacement.

This case ends tragically, if we use the term in its Classical sense, where it signifies not just death (real and/or symbolic), but also a separation, a tearing apart of the social fabric or its equivalent. There is the obvious, unfortunate end here to DJ's life story, a life story that is hardly a happy one by any definition, and that may be seen as tragic for any number of reasons; but that is only a part of what I mean by this case ending tragically. It is all the more tragic in that the surgeon chose not to adhere to his own values and best medical judgment, and which could, in fact, have been justified on the basis of doing little good, risking significant harm and in no way supporting this patient's interests and values, particularly insofar as these last were clearly unknown. Another tragedy is that the surgeon has lost his ethical compass. The ethical fabric, wherein he ought to function, and wherein he can function truly beneficently and wisely, has been torn because of his infidelity to his own values and medical judgment. Had he refused to perform the surgery, he would certainly have acted in a paternalistic manner. Some, perhaps many, would judge such an act to be one of strong paternalism, though I would not. I would assess a refusal to treat here as soft paternalism, for given the particular circumstances of this case a refusal to perform a valve replacement might have been the most beneficent act the surgeon could have performed. His insistence on the agreement as a condition for surgery was quite another thing, however. It was not only completely foolish in that it was unrealistic on its face, but it was an unconscionable violation of patient autonomy. Furthermore, inasmuch as this agreement would have had to be part of the informed consent process and document, it effectively rendered the surgical consent invalid by virtue of the coercion that it played in the process.

A Health Communication Scholar Responds

DJ's case resembles a case recently discussed by Hull and Jadbabai about a patient they named Mr. X (2014). Mr. X was an intravenous heroin user who presented with bacterial endocarditis requiring mitral valve replacement; his situation was exacerbated by multiple strokes and the fact that he signed himself out of the hospital against medical advice. Six months later he returned to the same hospital with severe sepsis and prosthetic fungal endocarditis, which required another surgical valve replacement. Three consultations with three different cardiothoracic surgeons agreed that Mr. X's continued injection drug use was a contra-indication for repeat valve replacement. Like DJ, Mr. X did ultimately have a second valve replacement (Mr. X needed a new mitral valve, DJ needed an aortic valve) and had a similarly difficult post-operative course and then died when his family implemented a do-not-resuscitate order. Unfortunately, the increased incidence of heroin use in the U.S. means that these ethically troubling cases may soon be ubiquitous.

Intravenous drug use is increasing rapidly in the United States. Injecting heroin has become the drug of choice among addicts since around 2012, when many states passed legislation regulating pain clinics and restricting the ability of physicians to dispense prescription pain medications from their offices (Ferraris and Sekela 2016). This legislation resulted in heroin becoming less expensive and easier to obtain than prescription opioids; heroin use almost doubled in the U.S. between 2006 and 2013 to 681,000 active users, with an estimated 169,000 starting use of the drug in 2013 (Substance Abuse and Mental Health Services Administration 2014). Heroin is now the most common illegal injected drug worldwide. The majority of people who inject drugs (PWID)[4] used to be young adults, although Wurcel and colleagues noted a recent trend toward a bimodal age distribution with peaks in the 21–35 year-old age group and the 46–60 year-old age group (Wurcel et al. 2016). The incidence of heroin-related deaths has surged as well, with nearly 6000 deaths in 2013, triple the number in 2006.[5]

The growing population of PWIDs in the U.S. has led to a growing number of people who are at risk for infective endocarditis that may require repeated surgical intervention. A recently released study of 436 patients undergoing surgery for active infective endocarditis revealed that 18% were current PWIDs (Kim et al.

2016). Overall the proportion of PWIDs who had infective endocarditis increased from 15% in 2002 to 26% in 2014. Kim and colleagues also found that PWIDs were younger and had fewer cardiovascular risk factors than people who did not inject drugs, but PWIDs had higher rates of valve-related complications, principally due to the higher rates of reinfection. Of the many medical complications of injection drug use, infective endocarditis is particularly challenging given the significant risk of operative mortality (estimated at between 8 and 37%), and social factors such as late recidivism and reinfection.

Patients who develop infective endocarditis due to intravenous drug use are not a favorite population for cardiothoracic surgeons to manage. They have a high recidivism rate, require intensive surgical interventions, and tend to rely on publicly funded medical insurance, which requires considerable resource expenditures on the part of acute care hospitals. Poor outcomes are related to drug resistance, delays in surgical treatment, the presence of concomitant risk factors and multiple organ dysfunction, acute congestive heart failure, prosthetic valve reinfection, and severity of valve injury (Ferraris and Sekela 2016; Kim et al. 2016; Wallace et al. 2002; Yamaguchi and Eishi 2007; Yankah et al. 2002).

Cardiothoracic surgeons, like the one in DJ's case, are frequently faced with the decision of whether they should perform repeat operations on a patient who has acquired infective endocarditis on a native or prosthetic valve resulting from continuing intravenous drug use. Surgeons are sharply divided on this issue (DiMaio et al. 2009). Some believe that patients should have one surgery, and if continued injection of drugs causes reinfection, the patient should only be offered antibiotic therapy. Surgeons in this camp believe there is no principled reason to do a second, or third surgery, and that spending time, energy and resources on a patient who has chosen to do him- or herself continued harm does not serve the greater good.

Others believe that claims of stewardship on the part of surgeons should be directed to the well-being of individual patients, rather than the distribution of medical resources overall. Surgeons on this side of the divide point to the fact that the primary problem for patients such as DJ is drug use, and that medical professionals must recognize drug addiction as a complex but potentially treatable disease. Asking patients like DJ to sign an agreement promising not to inject drugs, but offering no options for drug abuse treatment, counseling, or ongoing support is akin to punishment. Perhaps most persuasive is that one of the greatest risk factors for developing infective endocarditis is a previous heart operation for endocarditis, along with other factors such as poor dental hygiene. This certainly describes our patient: DJ developed infective endocarditis and had his first surgical valve replacement after he resorted to having dental work done in a garage (before which he probably should have had prophylactic antibiotics) (Nishimura et al. 2008). We do not know whether DJ was an active intravenous drug user when the second surgery was being considered. The used syringes in his hospital room and his level of lethargy unrelated to prescribed pain medication indicate the strong possibility of continued intravenous drug use, but the case narrative does not indicate whether any drug screens were conducted to confirm the health care team's suspicions. Recurrent endocarditis may result from problems unrelated to drug use and addiction, even if the patient remains engaged in injection drug use. Those in favor of repeated surgeries claim that surgeons unwilling to operate should transfer the patient to another surgeon or facility where the operation can take place. After all, physicians confront human frailty every day—lung cancer patients who cannot or will not stop smoking, obese patients who continue to overeat, or patients like DJ who may continue to inject drugs despite the negative consequences of doing so (which along with intense drug craving, defines addiction).

Surgeons on both sides of this divisive issue tend to focus on the immediate needs of the patient in front of them—should I operate (again) or not?—without necessarily considering the bigger picture. One could argue that DJ's problem was his injection drug use, not his heart. One thing surgeons could do is to insist that patients such as DJ meet with drug counselors and rehabilitation specialists while surgery is being considered. Patients with infective endocarditis spend 4–6 weeks in the hospital for intravenous antibiotic therapy, which provides ample time for surgeons and other physicians to gain their trust and at least introduce the idea of drug treatment along with surgery and other medical treatment and rehabilitation (Ferrari and Sekela 2016). The case narrative states that DJ did meet with a substance abuse counselor while he resided in a rehabilitation facility after his second valve replacement surgery. We know nothing about their conversation or what, if any, on-going support was offered, nor do we know whether DJ wanted to stop using drugs, or had in fact already done so. We do know that recovery from addiction requires a high level of on-going support, and that recovery may be interrupted with relapses, sometimes after long periods of abstinence (DiMaio et al. 2009).

Another thing surgeons who are concerned about the issue of repeated surgeries can do is to advocate for needle and syringe programs (NSPs) in their communities. Dirty needles and failure to disinfect the skin at the injection site greatly increase the risk of infection (Tookes et al. 2015). NSPs provide free sterile syringes and alcohol wipes, appropriately dispose of used syringes, and make a variety of health and supportive services, including on-site medical care, referrals for addiction treatment, screening and counseling for HIV, hepatitis C and other sexually transmitted infections available to injection drug users. They also may distribute condoms, food, and clothing, and provide referrals to other community resources. Such programs have been proven to reduce the spread of HIV and other serious infections, save money, encourage the safe disposal of syringes, minimize the risk of needle-stick injuries to law enforcement officials, and help link chemically dependent individuals to vital drug treatment services. Economic modeling of syringe exchange programs for prevention of HIV infection has found this strategy to be cost-effective and cost saving (Tookes et al. 2015). Given the substantial cost of severe bacterial and fungal infections and the need for repeated surgical intervention in many of these cases, estimates that consider only the cost and health impact of HIV and hepatitis C (although substantial in and of themselves) underestimate the potential benefits of these programs. For example, the annual estimated cost to operate a pilot NSP in Miami-Dade county was $202,451, less than the cost of treatment of only 6 injection drug users (estimating the median cost of treatment to be $39,896) (Tookes et al. 2015).

The Institute of Medicine has released data that show NSPs do not increase drug use, and participants in needle exchange programs are five times more likely to enter drug treatment programs (Hagan et al. 2000; Institute of Medicine 2006). There is also evidence that intravenous drug users are willing to participate in NSPs; at least one study found that 90% of syringes distributed through needle exchange programs were returned for safe disposal (Ksobiech 2004). Tookes and colleagues conducted visual inspection walkthroughs in a random sample of the top-quartile drug-affected neighborhoods in San Francisco (a city with needle and syringe exchange programs) and Miami (a city without such programs) and interviewed 600 injection drug users in San Francisco and 450 in Miami (Tookes et al. 2012). They found 44 syringes per 1000 census blocks in San Francisco and 371 syringes per 1000 census blocks in Miami. Thirteen percent of syringes injection drug users in San Francisco reported using in the 30 days prior to the study interviews were disposed of improperly, versus 95% of syringes in Miami.

The United States Congress approved the use of federal funds for needle and syringe programs in late 2009. The North American Syringe Exchange Network reported that as of 2015, there are approximately 200 needle-exchange programs in 33 states and the District of Columbia.[6] Harm reduction remains a controversial issue in medicine and politics. While cardiothoracic surgeons, because of their training, personalities and experience, may not see activism, prevention, or public health as within their primary domains of concern, they may be the most appropriate group to bring attention to the need to prevent, rather than react to, the projected epidemic level of surgical intervention needed in the coming years to treat infective endocarditis in people who inject heroin.

Notes

[1]Rabkin and colleagues reviewed the institutional policies of the Division of Cardiac Surgery at the University of Washington Medical Center, where their policy is to perform immediate surgery if patients have had septic cerebral emboli (like the patient Mr. X described in Hull and Jadbabai's case study; see #2 below) or active infection (like DJ). Had DJ been treated elsewhere, perhaps his surgery would not have been delayed, although Rabkin et al. concluded that it was unclear whether patients have better outcomes when they receive prompt surgical intervention in the setting of active infection. Their study also found a higher incidence of methicillin-resistant *Staphylococcus aureus* (MRSA) infection in intravenous drug users after valve replacement surgery, which was second only to older age at the time of surgery as independent predictors of diminished survival; DJ had both.

For more information see: Rabkin, David G., Nahush A. Mokadam, Donald W. Miller, Raymond R. Goetz, Edward E. Verrier, and Gabriel S. Aldea, G. S. 2012. Long-term outcome for the surgical treatment of infective endocarditis with a focus on intravenous drug users. *Annals of Thoracic Surgery* 93: 51–58.

[2]Justice is the last ethical principle typically applied to bioethical dilemmas concerned with patient treatment decisions. A host of pertinent issues such as allocation of scarce medical resources, the rights of persons to have access to health care, accelerating medical costs, etc. are often ignored when determining right action in the case of a particular patient. Health care reform remains a contentious issue in the United States, but ignoring issues pertaining to justice does not make problematic cases such as DJ's either less common nor easier to navigate.

For diverse accounts of justice in biomedical ethics, see Norman Daniels, *Just Health Care* (New York: Cambridge University Press, 1985); Daniels, *Just Health* (New York: Cambridge University Press, 2006); Madison Powers and Ruth Faden, *Social Justice: The Moral Foundations of Public Health and Health Policy* (New York: Oxford University Press, 2006); Allen Buchanan, "Health-Care Delivery and Resource Allocation," in *Medical Ethics*, ed. Robert Veatch, 2nd edition (Boston: Jones and Bartlett Publishers, 1997); Allen Buchanan, Dan Brock, Normal Daniels, and Daniel Wikler, From Change to Choice: Genetics and Justice (New York: Cambridge University Press, 2000); Kevin E. Hodges and Daniel P. Sulmasy, "Moral Status, Justice, and the Common Morality: Challenges for the Principlist Account of Moral Change," *Kennedy Institute of Ethics Journal* 23: 275–296.

[3]In examining the claim that patients have a "right" to an innovative surgical procedure that is yet to receive full FDA approval, Meyerson begins with the following assumptions: that all other known and/or reasonable therapeutic options have been tried; that patients agree to fund the costs entirely by themselves; that fully informed consent is obtained beforehand; that no coercion or subterfuge comes into play; and that Phase I trials of the new procedure have been completed, with no safety issues having been noted. See Meyerson, Denise. 2015. Is there a right to access innovative surgery? Bioethics 29: 342–352. Doi:https://doi.org/10.1111/bioe.12111.

[4]Terminlogy used to refer to the population of individuals who are injection drug users varies. I prefer the less judgmental PWID (people who inject drugs) favored by those in the harm reduction community. Articles published in medical journals tend to refer to either IVDUs (intravenous drug users) or the damning IVDAs (intravenous drug abusers). Similarly, heroin and other narcotics are often described as "illicit drugs." I agree with medical evidence that defines drug addiction as a disease and see nothing to be gained by using pejorative language to describe these patients.

[5]For more information see Centers for Disease Control and Prevention. Trends in drug-poisoning deaths involving opioid analgesics and heroin: United States, 1999–2012. Available at: http://www.cdc.gov/nchs/data/hestat/drug_poisoning/drug_poisoning.htm. Accessed October 21, 2016.

[6]See https://nasen.org/ for more information.

References

Bulger, Roger J., and John P. McGovern (eds.). 2001. *Physician philosopher: The philosophical foundation of medicine: Essays by Dr. Edmund Pellegrino*. Charlottesville, VA: Carden Jennings Publishing Co., Ltd.

Daniels, Norman. 1991. Duty to treat or right to refuse? *Hastings Center Report* 21: 36–46.

DiMaio, J.Michael, Thomas A. Salerno, Ron Bernstein, Katia Araujo, Marco Ricci, and Robert M. Sade. 2009. Ethical obligation of surgeons to noncompliant patients: Can a surgeon refuse to operate on an intravenous drug-abusing patient with recurrent aortic valve prosthesis infection? *Annals of Thoracic Surgery* 88: 1–8.

Ferraris, Victor A., and Michael E. Sekela. 2016. Missing the forest for the trees: The world around us and surgical treatment of endocarditis. *The Journal of Thoracic and Cardiovascular Surgery* 152: 677–680.

Gampel, Eric. 2006. Does professional autonomy protect medical futility judgments? *Bioethics* 20: 92–104.

Hagan, Holly, James P. McGough, Hanne Thiede, Sharon Hopkins, Jeffrey Duchin, and E. Russell Alexander. 2000. Reduced injection frequency and increased entry and retention in drug treatment associated with needle-exchange participation in Seattle drug injectors. *Journal of Substance Abuse Treatment* 19: 247–252.

Hull, Sarah C., and Farid Jadbabai. 2014. When is enough enough? The dilemma of valve replacement in a recidivist intravenous drug user. *Annals of Thoracic Surgery* 97: 1486–1487.

Institute of Medicine. 2006. *Preventing HIV infection among injecting drug users in high-risk countries. An assessment of the evidence*. Washington, D.C.: National Academies Press.

Kim, Joon B., Julius I. Ejiofor, Maroun Yammie, Masahiko Ando, Janice M. Camuso, Ian Youngster, Sandra B. Nelson, Arthur Y. Kim, Serguei I. Melnitchouk, James D. Rawn, Thomas E. MacGillivray, Lawrence H. Cohn, Joan G. Byrne, and Thoralf M. Sundt. 2016. Surgical outcomes of infective endocarditis among intravenous drug users. *The Journal of Thoracic and Cardiovascular Surgery* 152: 832–841. https://doi.org/10.1016/j.jtcvs.2016.02.072.

Ksobiech, K. 2004. Return rates for needle exchange programs: A common criticism answered. *Harm Reduction Journal*. http://www.harmreductionjournal.com/content/1/1/2.

Lepping, Peter, and Bevinahalli Nanjegowda Raveesh. 2014. Overvaluing autonomous decision-making. *British Journal of Psychiatry* 204: 1–2. https://doi.org/10.1192/bjp.bp.113.129833.

Luce, John. 1995. Physicians do not have a responsibility to provide futile or unreasonable care if a patient or family insists. *Critical Care Medicine* 23: 760–766.

Meyerson, Denise. 2015. Is there a right to access innovative surgery? *Bioethics* 29: 342–352. https://doi.org/10.1111/bioe.12111.

Nishimura, R.A., B.A. Carabello, D.P. Faxon, M.D. Freed, B.W. Lytle, P.T. O'Gara, R.A. O'Rourke, and P.M. Shah. 2008. ACC/AHA 2008 Guideline update on valvular heart disease. Focused update on infective endocarditis: A report of the American College of Cardiology/American Heart Association Task Force on practice guidelines endorsed by the Society of Cardiovascular Anesthesiologists, Society for Cardiovascular Angiography and Interventions, and Society of Thoracic Surgeons. *Journal of the American College of Cardiology* 52: 676–685. https://doi.org/10.1016/j.jacc.2008.05.008.

Pellegrino, Edmund D., and David C. Thomasma. 1988. *For the patient's good: The restoration of beneficence in healthcare*. New York: Oxford University Press.

Substance Abuse and Mental Health Services Administration. 2014. *Results from the 2013 National Survey on Drug Use and Health: Summary of national findings*. Rockville: Substance Abuse and Mental Health Services Administration. NSDUH Series H-48, HHS Publication No. SMA 14-4863.

Tookes, Hansel E., Alex H. Kral, Lynn D. Wenger, Gabriel A. Cardenas, Alexis N. Martinez, Recinda L. Sherman, Margaret Pereyra, David W. Forrest, Marlene LaLota, and Lisa R. Metsch. 2012. A comparison of syringe disposal practices among injection drug users in a city with versus a city without needle and syringe programs. *Drug and Alcohol Dependence* 123: 255–259.

Tookes, Hansel, Chanelle Diaz, Hua Li, Rafi Khalid, and Susanne Doblecki-Lewis. 2015. A cost analysis of hospitalizations for infections related to injection-drug use at a county safety-net hospital in Miami, Florida. *PLoS One* 10: e0129360. https://doi.org/10.1371/journal.pone. 0129360.

Wallace, S.M., B.I. Walton, R.K. Karbanda, et al. 2002. Mortality from infective endocarditis: Clinical predictors of outcome. *Heart* 88: 53–60.

Wreen, M. 2002. Medical futility and physician discretion. *Journal of Medical Ethics* 30: 275–278. https://doi.org/10.1136/jme.2002.000687.

Wurcel, Alysse G., Jordan E. Anderson, Kenneth K. Chui, K. Sally Skinner, Tamsin A. Knox, David R. Snydman, and Thomas J. Stopka. 2016. Increasing infectious endocarditis admissions among young people who inject drugs. *Open Forum Infectious Diseases.* Downloaded from http://ofid.oxfordjournals.org/ at University of South Florida on October 21, 2016.

Yamaguchi, Hiroichiro, and Kiyoyuki Eishi. 2007. Surgical treatment of active infective mitral valve endocarditis. *Annals of Thoracic and Cardiovascular Surgery* 13: 150–155.

Yankah, A.C., H. Klose, R. Petzine, et al. 2002. Surgical management of acute aortic root endocarditis with viable homograft: 13-year experience. *European Journal of Cardiac Surgery* 21: 260–267.

Case 16—A Depressed Caregiver Neglects His Own Health

JA was an 82-year-old man who was taken by ambulance to the local hospital in the middle of the night, having apparently suffered cardiac arrest. His wife, CA, had summoned help by repeatedly tapping on the floor of their 2nd story condominium with her cane, hoping to attract the neighbor's attention. CA was unable to communicate over the telephone, as she had suffered a devastating stroke in her late 30's, likely as a result of being a heavy smoker and an early adopter of birth control pills. When the neighbors finally heard her call for help, they found JA unconscious on the floor of the bedroom and quickly called 911.

Early the following morning, the neighbors found the name and local phone number of one of JA's brothers and called to alert him about what had happened. JA was the oldest of three brothers; the middle brother, who had received the neighbor's phone call, was spending the winter months locally with his wife. The neighbor told JA's brother what had happened and that an emergency placement in a nearby assisted living facility (ALF) was being arranged for CA by the hospital social worker. The brother called the hospital and was told that JA needed immediate quadruple bypass surgery, and his brother consented to the surgery on his behalf. JA did not have an advance directive nor had he specified a surrogate decision maker. CA was cognitively intact but her difficulties in communicating made her an unlikely surrogate.

JA's middle brother and his wife's daughter (JA's niece, who lived in the same state year- round) drove the 3 h to the hospital to check on him. JA was stable and heavily sedated after the surgery. The family visited CA at the assisted living facility, and found her distraught over her husband's condition and happy to get an update from her extended family. The brother and sister-in-law then visited the couple's condominium and were appalled by what they saw: old, spoiled food stored inappropriately in the dishwasher and in cupboards, dirty dishes spilling over the counters onto all available surfaces, and a urine-soaked sofa. The condominium was a disaster, and it was clear that JA's refusals to get together over the last few years should have been interpreted as calls for help. JA had been his wife's full-time

© Springer International Publishing AG 2017

L. A. Roscoe and D. P. Schenck, *Communication and Bioethics at the End of Life*, https://doi.org/10.1007/978-3-319-70920-8_16

caregiver for nearly 50 years. They had no children and no relatives nearby, and they had relied on one another exclusively.

JA's heart surgery was successful and he was finally able to admit to his brother that as a result of his wife's increasing disabilities, he had become unable to care for her and had become depressed as a result. "My greatest fear was that we would be separated," he explained, "that's why I didn't want anyone to see how we were living." JA underwent in-patient cardiac rehabilitation, and CA settled rather comfortably into her new assisted living residence. When JA was ready for discharge, the family worked with the hospital social worker and discharge planner to find a new assisted living facility where JA and CA could be together. Since CA could no longer swallow consistently enough to ensure adequate nutrition, the new ALF required placement of a feeding tube, to which she consented.

After a few relatively uneventful months, CA suffered another catastrophic stroke and died in the ambulance on the way to the hospital. JA's health continued to decline, exacerbated by his grieving. As is common after open-heart surgery, his depression worsened. He had been a life-long stutterer, which became more pronounced with his deepening depression and he also experienced the onset of mild cognitive impairment. JA's middle brother had also died in the interim, and so the youngest of the three brothers was appointed as JA's legal guardian. He lived in California, however, and told his niece in Florida that he trusted her to make whatever medical decisions were necessary for JA's care. JA's niece arranged for him to move closer to her family so that JA would not be so isolated after CA died. The niece also arranged for medical care for JA, assumed the role of health care proxy, and consulted with her other uncle in California only when necessary; the "California" uncle was also experiencing a number of serious medical challenges, and she did not want to add to his burdens. JA's niece was a professor at a major research university and had lived in the area for nearly 20 years. She had a number of contacts in the medical field, and arranged for JA to be a patient with a well-regarded geriatrician with whom she had collaborated on research in the past.

These bi-monthly doctor visits were something that JA and his niece looked forward to. They would rehearse answers to the questions JA knew he would be asked to assess his cognitive function such as "what day is it?" and "what did you have for breakfast today?" Often they would have a quick lunch together after JA's appointment. JA referred to the geriatrician as "Doctor," and even though his stutter and cognitive impairment made conversation difficult, the two nonetheless engaged in a kind of respectful dialogue, often punctuated with good-natured laughter. On one occasion the physician heard an odd heart sound and was relieved and amused to find a paper valentine heart pressed neatly inside JA's shirt pocket. A subsequent examination revealed the presence of cataracts in both eyes. One of his niece's first major responsibilities was to locate an appropriate ophthalmologist, schedule the surgeries, consent to the procedures, and accompany her uncle to the facility on two occasions. JA's improved eyesight contributed greatly to his quality of life. He enjoyed watching television with the other residents and was often found in the fenced backyard garden watching the birds and the squirrels. The new assisted living facility provided good care, served wonderful food, and created a warm and inviting atmosphere. Unlike many ALFs, this facility was part of a Medicare demonstration project that accepted patients' Social Security checks as sufficient payment for room, board and assistance with activities of daily living. JA participated in holiday parties and outings, and he was able to enjoy being around his niece and her young family.

As would be expected, however, his cognitive impairment increased, and his ability to participate in activities decreased as a result. JA's niece received a call from the ALF informing her that JA had been taken to a local hospital because he had begun to hallucinate and refused to eat. He was diagnosed with a urinary tract infection, which was effectively treated with antibiotics, and he was released after a few days. A few weeks later, the same situation occurred; this time the urologist said benign prostatic hypertrophy (BPH) was the likely cause of JA's repeated infections and occasional incontinence, and he recommended that JA undergo a transurethral resection of the prostate (TUR).

JA's niece consented to the surgery since JA was found to be incompetent by his attending physician. JA was capable of understanding what was happening to him, but it required an attentive listener to decipher was he was saying. His niece asked the urologist to speak to JA and obtain his assent to the procedure; the surgeon agreed to do so, but in fact did not take the time to explain anything to his patient. When his niece explained why he had to stay in the hospital, JA said, "I don't want surgery! Just let me go, I want to see my wife again!" His niece explained that the surgery was necessary in order for him to remain in the assisted living facility— were he to become permanently incontinent, he would have to be moved to a skilled nursing facility (SNF). The surgery proceeded, despite JA's protestations, and was successful. Upon discharge, JA returned to the ALF.

A follow-up appointment was scheduled for 3 weeks post-surgery. JA's niece picked him up at the ALF at 9:30 am. The ALF staff members were able to accommodate this scheduled appointment, and they made sure that JA was up and dressed and had had breakfast before his niece arrived. On the drive, JA became confused and attempted to get out of the car at a red light. Once they were safely at the medical building, the walk from the parking lot to the doctor's office took nearly 30 min. JA was disoriented and anxious, and he did not understand where he was going nor why. The urologist's waiting room was packed with 12 elderly male patients, most of whom were accompanied by a caregiver or relative. When his niece signed him in and asked about the likely waiting time, the receptionist informed her that all surgical follow-up appointments were scheduled for 11:00 am and that they would just have to wait until called. By about 12:30 pm, JA had become agitated, and was hungry and ready to return to the ALF for lunch and his usual afternoon nap. When he was finally called back, the nurse said there was no need for JA's niece to accompany him.

Several minutes later the nurse returned to the waiting room to summon JA's niece. She was red-faced and angry, and said JA refused to comply with her request to provide a urine sample. The nurse led his niece to a bathroom, and said loudly, "Sir, you need to pee in this cup immediately! And since you won't do it for me I brought your niece, and she is going to have to make you do what I say!" JA looked ashamed; he had no idea what he was expected to do and lacked the ability to comply with the nurse's impatient demands. JA's niece explained that if a urine sample was required, she would have had to make arrangements with the ALF to procure it the day before. JA was incontinent and too demented to be able to produce a urine specimen on demand, especially in a strange and threatening environment. The nurse, clearly annoyed, explained that the follow up visit would be a waste of the doctor's time and that they might as well leave. JA was bewildered by the commotion and disruption to his daily routine, and his niece was frustrated and concerned that her uncle was not receiving the post-surgical care he required.

A month later, the ALF called JA's niece to say he had been taken to the hospital by ambulance because he had vomited blood. The ER physicians diagnosed him as having diverticulitis.[1] JA's treatment involved keeping the stomach empty by sucking out the contents through a tube passed up his nose and down his throat into his stomach (nasogastric or NG tube). JA was thus able to avoid surgery, but his hospital stay proved difficult. He again experienced hallucinations, and was extremely confused about where he was. When his niece visited he explained that he needed to get up to feed the cats who he said were crouched under his hospital bed. JA had been a life-long animal lover, and a favorite family story was about the winter in Michigan when JA had fed 20 stray cats he found huddled in an abandoned warehouse. He became frightened when anyone came near him, and he eventually needed to be restrained to prevent him from pulling out his NG tube or IV. The restraints increased his terror and confusion, and his niece intervened. Instead of requiring physical restraints, the hospital installed a "watcher," a volunteer who sat quietly in his room in order to provide a comforting presence. After discharge to the assisted living facility, JA told his niece that he did not ever want to return to the hospital. The next time the ALF called to report that JA had once again been taken to the hospital because of hallucinations and a probable infection of some sort, his niece requested a hospice evaluation; the hospital did not have a palliative care service.

A nurse from the local hospice agency came to JA's hospital room to evaluate his eligibility for hospice services. She said that although JA did not have a terminal or end-stage disease, he was hospice-eligible under the general diagnosis of "failure to thrive." He was thus able to return to the ALF under the care of hospice upon discharge from the hospital. A hospice nurse visited him twice each week at the ALF. The hospice nurse was also able to spend time with the certified nursing assistants who provided most of the care at the ALF, and as a result, the care of all residents, including JA, was enhanced. After 3 months JA's case was re-evaluated by the hospice care team, and because of his improved mood and functional status, JA was found to no longer meet the hospice criterion of "expected death within 6 months," which was required in order for the ALF to be reimbursed for JA's care through the Medicare Hospice Benefit.

After the hospice nurse's visits stopped, JA's condition predictably deteriorated. His niece requested another hospice evaluation after 6 weeks, and JA was readmitted to hospice care. His condition improved to the point at which the hospice found him ineligible after another 3 months. His condition deteriorated again, and the ALF administrator sent him to the local hospital when his aides were unable to get him up and dressed for breakfast. By now JA was incontinent, moderately cognitively impaired and unable to effectively communicate. When the hospital was not able to diagnose any condition requiring in-patient acute care, JA was discharged to a skilled nursing facility, since the ALF was no longer willing to accommodate his incontinence and repeated needs for hospitalization.

JA spent 2 weeks in the nursing home and was then admitted to the hospital due to his inability to ingest sufficient calories by mouth. His niece met with the attending physician, who asked if she would consent to feeding tube placement. JA was only intermittently conscious and unable to communicate and had been diagnosed with bilateral pneumonia. Knowing her uncle's wishes for no further hospitalizations, his niece refused to provide consent for the feeding tube or for antibiotics. JA died peacefully two days later, with his niece at his bedside.

Discussion Questions

1. Under what circumstances should a proxy's decision making authority be questioned?
2. What are some ways to include incapacitated patients in decision making?
3. How can caregivers be supported so as not to neglect their own health?
4. In what ways does the Medicare Hospice Benefit limit hospice agencies from being able to fulfill their mission of providing high quality end-of-life care?

A Bioethicist Responds

The case of JA appears to be similar to so many others involving elderly patients who have become incapacitated, who have left no advance directives and who have not designated specific surrogate decision-makers. Yet consideration of four specific issues will yield a rather interesting analysis of this situation:

1. The presumed "authority" of JA's niece as health care proxy;
2. The role of his niece as advocate (proxy) for JA;
3. The urologist's office environment, including staff;
4. Prudential decision making on the part of JA's niece.

The case narrative makes clear that JA's niece assumed total responsibility for her uncle's health care, effectively becoming his proxy, once JA's wife and his middle brother are deceased, and once JA's youngest brother in California, who had been appointed his legal guardian, has entrusted "whatever medical decisions were necessary" into her care. Yet this immediately raises an important issue: There is no evidence that his niece had been legally appointed as health care proxy, or that this authority should fall to her automatically as provided in Florida law. F.S. 765.401 lists in priority order those persons who may act as proxies for incapacitated or developmentally disabled patients who have not executed advance directives or designated surrogates to execute an advance directive.[2] Nowhere in that statute does it say anything about nieces or nephews. However, part of this statute does provide that "An adult relative of the patient who has exhibited special care and concern for the patient and who has maintained regular contact with the patient and who is familiar with the patient's activities, health, and religious or moral beliefs;..." may act as health care proxy or surrogate. One might argue that JA's niece exhibited special care and concern for her uncle as well as familiarity with his activities, health and religious or moral beliefs. Nonetheless, his niece picked up the mantle of proxy by virtue of JA's youngest brother having merely handed it over to her absent any formal legal action or documentation. And, she is apparently never questioned by anyone charged with the responsibility for providing health care for her uncle as to whether or not she has authority to act as proxy.

It may nevertheless be unsurprising that JA's niece has little trouble stepping into this role unchallenged. She had arranged to have her uncle moved closer to her and her family so that he not be isolated after the death of his wife, and no doubt those familiar with JA and his health problems at that time were only too happy to have someone with initiative and the apparent resources come to his rescue, and move him from their area of responsibility. It can also safely be assumed that JA's niece would have introduced herself to all health care, hospital, physician's office, ALF, SNF and insurance office personnel as "Dr. X", as would be customary for a university professor in dealing with others in a business or professional setting. It is not to suggest that JA's niece would in any way have attempted to masquerade as a physician in order to somehow intimidate persons accustomed never to question the authority, statements or actions of those addressed as "Doctor," but the foregoing is to suggest that her professionalism, self-possession and willingness to assume the advocacy role, especially in the absence of challenges to her doing so, goes a long way to explaining the health care community's apparent ready acceptance of the patient's niece as legitimate proxy without insisting upon documentary evidence.

Her continuing concern and attention to her uncle's needs served only to underscore her good intentions and wise choices surrounding his overall health care. This was demonstrated not only by her choice of ALF, but also her guarantee of proper geriatric, ophthalmological and urological care for him, signing consents as necessary. She would later also ensure that a hospice nurse visit him at his ALF. JA's niece clearly filled the role of the advocate with the best interests of the patient in mind and, therefore, acting as health care proxy appropriately. It was not all easy going, however.

JA's condition deteriorated somewhat rapidly after the TUR, particularly with regard to confusion and dementia, as is seen most dramatically in the three-week post-operative office visit to the urologist. It is here that his niece experiences the worst of the medical establishment's elitist mentality or behavior that can some-times be used to intimidate patients and let them know who the most important, the most central figure in the healing relationship really is. This urologist's office had done what many surgical specialists have been known to do, that is schedule all post-operative office follow-ups for the same hour, thus ensuring that a queue of patients will be ready to be ushered into the next available examining room as soon as one has been vacated such that the physician can be assured of having absolutely no delay in moving from one patient to the next—at her or his own speed. The fact that this may require some patients to wait virtually hours to be seen, not to mention that this type of scheduling is nothing short of misleading, says a great deal about who the physician believes is the more important in the physician-patient dyad, whose time is more valuable, and who is in "control" in this relationship. None of this could have been lost on JA's niece, who had to have already been frustrated in just getting her uncle to the doctor's office that day, but the nurse's behavior, after attempting to obtain a urine specimen from JA, only added insult to injury. Such treatment of patients or family members by any professional is unconscionable.[3]

The resolution to what essentially becomes a hopeless situation for JA occurred when his niece refused consent to feeding tube placement, and refused antibiotics on her uncle's behalf when he contracted pneumonia. She had now made a health care decision that she believed was in her uncle's best interest, given that he had become semi-conscious and unable to communicate, was unable to take in adequate nutrition by mouth, had expressed to her his unwillingness to endure further medical interventions, and had been going through a revolving door between the SNF and the hospital. There are undoubtedly those who would argue that she effectively abandoned her uncle, withheld life-sustaining care and thus "killed" him, albeit a view the present writer cannot share.[4] In any case, it had become increasingly clear that JA's progressive decline was unlikely to be reversed. He could be seen to appreciate virtually no quality of life at that point. Moreover, it must be remembered that he had said he was ready to die so that he could be reunited with CA, his wife. It would appear from what is known of him that JA has experienced what he wanted and what he could expect from life at that point, that he viewed his life story as complete, and that his niece would be doing more harm than good by forcing additional treatment upon him. It was not the withdrawal of treatment that brought about JA's death; it was rather the natural result of the disease process. We may assert, therefore, that his niece made a careful assessment of his situation and took the prudential decision that yielded the most desirable result for this particular patient, while also producing the greatest amount of good along with the least amount of harm. A difficult decision to be sure, particularly insofar as it must be made for a family member, but there was nothing in the case narrative to suggest that there was any ulterior motive or conflict of interest on the part of JA's niece.[5] There is every reason to believe that no matter how difficult it would be for anyone in her position to maintain any sense of objectivity in such a situation, she had succeeded in employing the requisite measure of *phronesis* in order to resolve her uncle's situation ethically.[6]

A Health Communication Scholar Responds

This is, unfortunately, a fairly typical story. It lacks some of the emergent ethical issues of previous cases, but is included here to highlight the unique vulnerabilities of elderly male caregivers, the despair felt by those in failing health whose health status is poor but who are not yet fully eligible for hospice care, and the steady resistance to advance care planning that occurs in many families.

Unlike the caregiving relationship between JA and CA, but characteristic of the later caregiving relationships between JA and his niece, most caregiving falls to women. Daughters, wives, mothers, sisters, daughters-in-law, and other female family members provide most of the world's caregiving to the vulnerable and needy members of the family and the community. But the number of men caring for an older adult has doubled in the past 15 years, from 19% of caregivers in 1996 to 40% by 2009, according to data from the Alzheimer's Association[7] and the National Alliance for Caregiving (NAC).[8] Men face different obstacles and stressors in providing care than do women.[9] Sheer inexperience can raise stress levels. Men are less prepared for the caregiving role, and have less experience in dealing with problems like incontinence, bathing, or feeding, and less practice managing household tasks like cleaning and food preparation. Men tend to be less likely to ask for help, or to be embedded in social networks that provide support such as a shoulder to cry on, information about available resources, or instrumental support like help with meals or transportation. Men who are married tend to rely on their spouse to remind them about important self- and preventive care—regular doctor and dentist appointments, a healthy diet, adherence to medication regimens—and may be unfamiliar or uneasy taking on these health maintenance tasks for themselves and others. Nearly three quarters (72%) of family caregivers report not going to the doctor as often as they should and 55% say they skip doctor appointments for themselves.[10] Men have also traditionally been socialized to keep their emotions to themselves, and tend to view mental health problems like depression or anxiety as personal failings rather than medical conditions, at least among older cohorts.

Depression can be a serious side effect of prolonged caregiving among both men and women. It is estimated that 40–70% of family caregivers have clinically significant symptoms of depression and approximately a quarter to half of these caregivers meet the diagnostic criteria for major depression (Cohen 2000). An older man whose wife is dependent on him for care may develop serious depression after years of caregiving, coupled with increasing isolation. Heart disease and cardiac surgery also increase the incidence and severity of depression; JA was both a long-time caregiver and a cardiac patient (Malphurs and Roscoe 2001). In extreme instances, depression and isolation produce hopelessness in the male caregiver and may trigger acts as desperate as homicide-suicide: murdering one's dependent spouse, and then taking one's own life. An analysis of homicide-suicides in West Central Florida revealed that approximately 40% of the perpetrators (male caregivers) had depression or other psychiatric problems (Cohen 2000). A common feature that precipitates these violent acts is a perception by the older man of an unacceptable threat to the integrity of the relationship (such as impending institutionalization), or a real or perceived threat to the caregiver's health. The present situation did not include violence, but clearly demonstrates to what extremes an older caregiver may go to maintain the current caregiving arrangement, even if it meant ignoring one's own health status, accepting a less than ideal standard of living, and refusing to accept the interference of even the most well-meaning family members.

JA's situation also highlights the dilemma faced by patients who truly do not want medical interventions that will prolong their lives, but who are not close enough to death to qualify for hospice care without interruption. JA did not perceive he had an acceptable quality of life, even though his basic needs were accommodated. He had difficulty communicating and keenly missed the intimate companionship he enjoyed with his wife of so many years. Because he was unable to meet the standards of capacity for informed consent in the hospitals providing his care, his niece was put in the uneasy position of forcing her uncle to allow procedures that he did not want to undergo. The uncomfortable truth here is that we do not have a health care system that can honor such a request. If a patient is receiving life-sustaining procedures like ventilators or feeding tubes and wants them discontinued, then either the patient or his or her surrogate can make the decision to withdraw these life-supportive measures and allow the patient to die. If a patient is believed to be within 6 months of dying, then hospice care provides a means of providing comfort and support during the dying process while forgoing procedures that might prolong life.

Unfortunately, the lines between terminally and seriously ill are not always easy to draw, and institutional requirements dictate certain rules and regulations be followed. JA did not want the TUR, nor did he want to be hospitalized for treatment for diverticulitis. However, few if any assisted living facilities or nursing homes are willing to care for a patient who is hallucinating due to repeated urinary tract infections, nor one who is at risk of bleeding out because of complications from diverticulitis. The indicated procedures had to be performed if JA was to continue to sustain the quality of life he enjoyed as a resident of the assisted living facility. Returning to the nursing home would have required feeding tube placement along with antibiotic therapy.

Hospice care was deemed appropriate, not appropriate, appropriate yet again— also due to the institutional regulations that govern the use of the Medicare Hospice Benefit. The hospice that provided JA's care had recently been audited by Medicare and had been accused and fined for having too many patients who outlived a 6-month journey to death. And indeed, JA fit the profile of just such a patient— someone whose physical and mental status improved rather dramatically with the addition of skilled hospice nursing care, which addressed pain and symptom management and quality of life specifically. Once he improved to the point of being perceived as no longer having a terminal "failure to thrive" prognosis, those services were removed, and his health declined. This is a deplorable situation. One must assume that if patients who are close to death and who do not want to have aggressive or really any curative medical treatment will improve if their pain and other symptoms are expertly addressed. Such patients deserve to receive hospice benefits no matter if their prognosis exceeds an anticipated 6-month time frame. In several instances JA's hospice eligibility was terminated before having reached the 6-month time line.

Other evidence of the need to revise Medicare Hospice Benefit guidelines can be found in studies that examine the factors that encouraged Jack Kevorkian's clients to seek his aid in dying. Kevorkian was a retired pathologist who assisted in the deaths of over 100 people in Oakland County, Michigan between 1990 and 1997 (Roscoe et al. 2000, 2001). Kevorkian provided illegal assistance in dying to patients who responded to his advertisements in a local newspaper and his growing infamy. The majority of his clients—75%—were not seen as terminally ill (i.e., within 6 months of dying) and were thus unable to access hospice care. In fact, only 3% of the 69 cases included in this study were under the care of hospice. For the most part, these men and women fell into a black hole in our medical care system: too sick to truly benefit from additional surgeries, rounds of chemotherapy, or other potentially curative treatments, and too well to be admitted to hospice care. In their extreme desperation, they sought Kevorkian's illegal help in ending their lives, either with the aid of his intravenous "suicide machine," or if lethal drugs were not available, through the time-tested method of carbon monoxide inhalation. Kevorkian eventually, after 4 jury trials and the enactment of a Michigan law that specifically forbade his acts of assistance in dying, was convicted of second degree murder and served prison time; he died shortly after his release.

The final issue to consider in the case of JA concerns reluctance to engage in advance care planning. The rates of completion of advance directives continue to be low. It will be interesting to see if the new billing codes for conversations and documentation of end-of-life preferences will result in higher rates of completion.[11] We hope that our family members will look out for our medical and other needs, and will make decisions that best represent what we would have wanted if we were able to make our own decisions. It is usually easier for family members to make decisions for their loved ones if they have some documented evidence of their treatment preferences; doubtless JA's niece would have felt a slight bit less alone in making the decision to withhold antibiotics during his last hospitalization if her uncle's advance directive had directed her to do so. Even though they were not documented in an advance directive, JA's preferences were clear to his niece, and perhaps clear to her alone. It was fortuitous in some ways that the hospitals and doctors who cared for JA were so lax in insuring that she had either the legal right to represent his decisions, or had his best interests at heart. At its best, the notion of advanced care planning would encompass more than preferences for medical treatment. Conversations between family members and with physicians should also investigate the preferences and plans older adults have for their living situations, financial interests, and quality of life overall, especially since in many cases prolonged needs for care and multiple medical decisions are likely.

Notes

[1]Diverticulosis occurs when pouches (diverticula) form in the wall of the colon. If these pouches get inflamed or infected, the condition is called diverticulitis, which can be very painful.

[2]The following rank-ordered list is provided on F.S. 765.401 (abbreviated here):

(a) The judicially appointed guardian of the patient of the guardian advocate of the person having a developmental disability...;

(b) The patient's spouse;

(c) An adult child of the patient, or if the patient has more than one adult child, a majority of the adult children who are reasonably available for consultation;

(d) A parent of the patient;

(e) The adult sibling of the patient, or if the patient has more than one adult sibling, a majority of the adult siblings who are reasonably available for consultation;

(f) An adult relative of the patient who has exhibited special care and concern for the patient and who has maintained regular contact with the patient and who is familiar with the patient's activities, health, and religious or moral beliefs; or

(g) A close friend of the patient.

(h) A clinical social worker licensed pursuant to chapter 491, or who is a graduate of a court-approved guardianship program. Such a proxy must be selected by the provider's bioethics committee and must not be employed by the provider.

[3]In fairness, there are two sides to this issue: optimizing effective use of the physician's time with each patient while minimizing patient waiting time. There is considerable literature on the issue of office and clinic scheduling, one of the oldest and most well-known of which is the Bailey-Welch rule of double-booking the first appointment slot and then assigning a single patient to each succeeding slot. See Bailey, Norman. 1952. A study of queues and appointment systems in hospital outpatient departments with special reference to waiting times. *Journal of the Royal Statistical Society* 14: 185–199; and Welch, J. D. 1964. Appointment systems in hospital outpatient departments. *Operational Research Quarterly* 15: 224–232. More recent articles discuss systems derived from mathematical and/or statistical models that account for variables not considered by Bailey and Welch (e.g., low complexity cases, high complexity cases, patient time with nurse, patient time with physician). See especially Oh, H. J., et al. 2013. Guidelines for scheduling in primary care under different patient types and stochastic nurse and provider service times. *IIE Transactions on Healthcare Systems Engineering* 3: 263–279. The practice of scheduling all post-op visits at the same time, however, is quite another thing.

[4]This debate will not be taken up here. Readers interested in delving into this subject will find a wealth of information in any good bioethics textbook, or in journals such as *Bioethics, Journal of Medical Ethics, The Journal of Clinical Ethics, The Journal of Medicine and Philosophy, Cambridge Quarterly of Healthcare Ethics, Kennedy Institute of Ethics Journal, American Journal of Bioethics* and *American Journal of Hospice and Palliative Medicine*, among others.

[5]To clarify, JA's niece did not inherit anything from her uncle upon his death. His condominium, car and other assets were left to his younger brother in payment for his service as executor of the estate.

[6]One of the best treatments of phronesis in the field of biomedical ethics can be found in Pellegrino, Edmund D., & David C. Thomasma. 1993. *Virtues in medical practice*. New York: Oxford University Press.

[7]Refer to Alzheimer's Association (www.alz.org) for more information.

[8]See National Alliance for Caregiving, Caregiving in the U.S. 2015. (www.caregiving.org/caregiving2015/)

[9]For more information see Scott, P. S. Caregiver stress syndrome: What's different for men. (https://www.caring.com/articles/caregiver-stress-syndrome-different-for-men)

[10]Refer to Caregiver Action Network (http://caregiveraction.org/resources/caregiver-statistics)

[11]For more information, consult the Centers for Medicare and Medicaid Services (CMS) website: https://www.cms.gov/Newsroom/MediaReleaseDatabase/Press-releases/2015-Press-releases-items/2015-10-30.html

Two Medicare billing codes were added in 2015 to allow physicians to bill Medicare for conversations about end-of-life care preferences, and can include the completion of advance directives or any other relevant legal forms (Living Wills, Health Care Proxy, Health Care Durable Power of Attorney, and Medical Orders for Life Sustaining Treatment) if applicable.

References

Cohen, Donna. 2000. Homicide-suicide in older people. *Psychiatric Times* XVII: 1–7.

Malphurs, Julie E., and Lori A. Roscoe. 2001. Neurocognitive function after coronary-artery bypass surgery. *New England Journal of Medicine* 345: 544 (Letter to the Editor).

Roscoe, Lori A., Julie E. Malphurs, L.J. Dragovic, and Donna Cohen. 2000. Dr. Kevorkian and euthanasia cases in Oakland County, Michigan, 1990–1998. *New England Journal of Medicine* 34: 1735–1736.

Roscoe, Lori A., Julie E. Malphurs, L. J. Dragovic, and Donna Cohen. 2001. A comparison of Kevorkian euthanasia cases and physician-assisted suicides in Oregon. *The Gerontologist* 41: 439–446.

Conclusion

> The responsibility toward other people's stories is real and inescapable, but that doesn't mean appropriation is the way to satisfy that responsibility. In fact, the opposite is true: Telling the stories in which we are complicit outsiders has to be done with imagination and skepticism.
>
> —Teju Cole

We have indeed presented the cases here as "complicit outsiders," and have endeavored to do justice to the situations we were privileged to witness and in some cases, participate in attempting to resolve. We have brought both our imaginative musings and a degree of skepticism that the difficult dilemmas in this book can be successfully resolved. It is our hope that by enlisting the efforts of others who would also feel complicit in how hospitalized patients die in the U.S., we can work toward solutions that are more inclusive, compassionate, and dignified for all.

The cases included in this volume presented many situations in which competing ideas about the "right thing to do" arose when the stakes were high. We live in an historical moment characterized by differing ideas of the good in almost all facets of life: education, politics, family life, and health care to name but a few. Gone are the days of tolerance of paternalistic physicians, or even agreement on where one should look for guidance in making end-of-life decisions. We have no shared "common sense" anymore about how life is to be lived, or how death can best be achieved (Arnett et al. 2009, 64). In such a time, the most important ethical commitment is to learn about competing ideas of the good. Most of us are far better at explaining our positions and trying to persuade others than we are at really listening and attempting to find at least a minimal set of ideas on which we might agree despite our differences (Arnett et al. 2009; Benhabib and Dallmayr 1990). We live in a time when learning and listening are more important than persuading and judging. Our hope is that this casebook has provided its readers with opportunities to learn more about different perspectives on challenging situations. Commitment to an ongoing "moral conversation" is more important than the desire to achieve consensus by persuading others to agree with what we think is the right course of action (Benhabib and Dallmayr 1990, 346).

As we present our own view of the themes that best characterize the compelling ethical dilemmas presented in this book, five major issues have surfaced

© Springer International Publishing AG 2017

181

L. A. Roscoe and D. P. Schenck, *Communication and Bioethics at the End of Life*,
https://doi.org/10.1007/978-3-319-70920-8

prominently: poor access to health care; the intersection of cultures/religions; the "validity" of informed consent; problems with family members; and problems with hierarchy and teamwork.[1]

PoorAccess to Health Care

There have been a variety of legislative efforts over the past twenty years to improve access to health care in the United States. All of these contained significant flaws and were ethically controversial. For example, legislation in 2010 moved the country closer to achieving universal health care, but costs have continued to rise and nearly 26 million Americans are still uninsured according to the Congressional Budget Office. As this volume goes to press, further efforts are underway to revise or eliminate the gains made by this previous legislation. Case 1 (about CS, a pregnant prostitute dying of head and neck cancer) and Case 15 (about DJ, an unfunded patient who needed repeated heart valve replacements because of continued drug use) both highlight this issue. CS's unfunded status and social situation, undoubtedly complicated by her image as "the pregnant prostitute," reveal what can happen to someone without regular access to health care despite the fact that she was fortunate enough to receive excellent surgical and maternity care prior to death by virtue of a public safety net. The situation surrounding DJ was even more complicated for he was not only unfunded but burdened with a complicated medical history and a serious drug addiction, which he was unable to control and which led to a series of infections, while he attempted to bargain with his cardiothoracic surgeon for a repeat heart valve replacement. In his case the *scarce resource problem* became coupled with that of *access to care* as well as the question of a *right to care.*

It appears to us that there is more interest in investing in cutting edge medical technologies that might benefit a select few patients than there is in achieving broad access topreventive care such as prenatal care and childhood vaccinations, and similar services for adults designed to reduce the incidence of diabetes and heart disease. A recent report from the Commonwealth Fund[2] found that the U.S. trailed other developed countries in making medical care affordable and ranked poorly in providing timely access to medical care. Yet Americans spend $9523 per person per year on medical expenses—by far the most among developed countries—and still life expectancy in the U.S. is much lower. Many of us tend to credit the enormous gains in life expectancy over the past century to the rise in medical technology. Surely technology and pharmacology play a role, but the improvements in the

---

[1]We do not attempt to place each of the 16 cases in this book within one of the five major groups we have identified; any attempt to do so would be to force the issue. Similarly, readers should understand that a particular case might well be included in more than one group.

[2]David. Squires and C. Anderson, *U.S. Health Care from a Global Perspective: Spending, Use of Services, Prices, and Health in 13 Countries,* The Commonwealth Fund, October 2015.

health of our population due to cleaner water and air, better workplace safety efforts, and preventing maternal and infant deaths should not be overlooked. Similarly, increasing access to even basic health care would likely do much to improve the overall health of our population.

Intersection of Cultures/Religions

Several situations were featured where non-western cultures contrasted sharply, or conflicted, with generally accepted cultural norms, health care practices, laws and/or traditional values of the west; or, where a patient, firmly committed to a particular western religious practice, came into seemingly irreconcilable conflict with the usual norms and practices of American health care. Physicians, nurses and other American health care professionals can face enormous challenges in dealing with persons native to China or other Asian cultures. Case 2 described the dilemma of a young Chinese couple grappling with the needs of their Down syndrome baby. While this case raised a number of ethical issues, the fact that it had to do with a Down syndrome baby born to Chinese graduate students studying in the U.S., that is, to parents who did not desire treatment for an infant they viewed as "defective," underscored stark cultural differences between the Chinese and American people, not only in terms of traditional values but also religious beliefs and socio-cultural attitudes towards certain groups of persons. Cultural differences between east and west can be even more challenging as witnessed in the tragic events surrounding the death of LF in Case 11, whose Chinese husband beat her nearly to death yet continued to serve as her health care proxy. It may seem impossible at times to gain anything close to a full understanding of the mindsets of persons from cultures with which one has little or no first-hand knowledge and experience (including such things as linguistic, social, religious, political and historical factors). However, the case of VH (Case 8) did reveal how an open, empathic and compassionate health care team made every effort to understand and accommodate a Hindu patient and her extended family.

Two other cases should be singled out for inclusion in this category: first, the difficult case of MT in Case 12, born with Turner Syndrome, who required aortic valve replacement at age twenty-five, had nothing to do with a foreign culture, but rather with a Christian denomination. We highlight this case because of the conflict arising from the specific instructions MT had written into her operative informed consent for surgery reflecting religious prohibitions of Jehovah's Witnesses, as well as the intense pressure brought to bear on the surgeon and his team by MT's future mother-in-law and an Elder from the local Jehovah's Witness Kingdom Hall. The second one deals with another western culture, which happens not to be all that "foreign" to many health care settings in the United States, specifically, an American neighbor to the south. Case 13 documents the situation of thirty-nine year-old JG, an undocumented Mexican who had lived and worked in the U.S. for fifteen years and had two children (ages eight and ten) with his significant other. Neither JG nor his partner had insurance coverage or were eligible for government aid, and neither had an advance directive or had identified one another as their health care surrogates. JG ended up in a persistent vegetative state after a motor vehicle accident, and he became the unfortunate, unwitting ward of the hospital.

The "Validity" of Informed Consent

Adults who are competent have the right to make their own health care decisions, which is protected by the right to privacy guaranteed by the Constitution of the United States. This right is grounded in the philosophical principle of autonomy, and the practical application of this principle is the practice of informed consent (Gracia 2012). There are two issues related to informed consent in this volume: The *interpretation* of informed consent, and what may actually constitute *full disclosure* in the consenting of a patient. Situations in which the interpretation of informed consent is involved may include how a patient's consent, written, stated or implied, may be loosely followed or even directly contradicted for some ostensibly good reason (particularly when the once-competent patient may no longer have capacity or may no longer be living); a patient's "self-contradiction" of what he clearly would like for himself; or someone acting as surrogate for an incapacitated relative but where this acting surrogate has, in fact, no real authority to exercise this formal, legal role.

The fascinating situation in Case 5 where MB offers her late husband's cryopreserved sperm to her sister, LC, is perhaps the most dramatic instance of an *interpretation* of informed consent that we have examined. Alongside the ethical issues of this case discussed earlier may now be added the fact that although MB provided a valid, legal consent for post-mortem sperm retrieval, she did so at the time with every intention that it one day be used for her own pregnancy. Without reviewing again the ethical questions raised in her ultimate donation of this sperm, to whom and for what purpose, her decision regarding the donation can only be viewed as one based upon an *interpretation*, or a "reformulation," of the consent she had given earlier; while her ultimate disposition of her late husband's sperm is very different from what had been her original intentions, her later actions do not appear to contravene the honest intentions of her consent to sperm retrieval in view of the circumstances of this case.

We witnessed an even broader *interpretation* of informed consent in Case 16, where JA's niece assumed the role of health care proxy for her uncle when his last remaining family member and legal guardian, a brother in California, told her he trusted her to make whatever medical decisions necessary for JA's care. She consulted with this brother in California only when necessary, and gradually took on full decision-making responsibility for JA despite having no real authority to do so. This ultimately included refusing consent to feeding tube placement and antibiotics for pneumonia, all as proxy on behalf of JA, at the end of his life. We saw earlier how all this managed to occur without challenge, yet it remains a rather striking example of how broad an *interpretation* of consent can be.

Advance directives were developed to extend a patient's decision making authority beyond a time when he or she might be competent to voice such preferences; done well, advance directives effectively provide "consent" from an incompetent patient near death for certain kinds of medical care. In Case 9, where an adult patient's mother contests the validity of her adult daughter's advance directive, provides an excellent example of how fragile such a document can be in the face of a parent or other family member who has decided to insist on what he or she feels is best for their loved one, in spite of evidence to the contrary.

Informed consent also raises questions related to *full disclosure*, which was underscored in Case 14. As mentioned in the case narrative, there was no consensus among those who attended tumor board conference the day CB's case was discussed; some members of the team felt CB should have the pros and cons of three treatment options fully laid out for him, after which he and the attending physician could make an informed choice; others felt that because of serious, complicating factors it would be inappropriate to offer him all options and that palliative care should be the only option offered. We also recall that the surgeon decided after private reflection to "play it by the book" and offer CB the three options, even though she did not feel all were advisable. His decision for palliative care not only surprised his surgeon but stands as a further lesson of the value of full and thorough disclosure in the process of informed consent.

Full disclosure requires not only that patients understand the risks and benefits of the proposed or recommended treatment, but also the benefits and potential burdens of all available options, including the option to cease aggressive treatment in favor of palliative care or comfort measures only. In our experience, too often patients are given only the information they need to make the *next* decision, which deprives patients of the ability to determine the route they may have wanted to take from the point of diagnosis until death (Roscoe et al. 2013, 190). Each decision a patient makes sets them along a particular trajectory, and as many of the cases in this book illustrate, it can be difficult for all involved to move away from increasingly aggressive and futile treatment until all options are exhausted, even those with extremely small odds of success.

Problems with Family Members

Almost every case included in this book includes examples of the ways in which family members, often with the best of intentions (but not always) are able to subvert the wishes of the patient, question the knowledge of the physicians, and try the patience and compassion of every nurse, social worker, or administrator involved in the patient's care. There are many reasons and explanations for why family members act and react in the ways they do when confronted with the serious medical condition of a loved one. Most families are unprepared for such difficult situations, and most have not had conversations that would allow them to help their family member make the decisions necessary to insure good care. Many people, even those with advanced educations or prestigious careers, lack health literacy, i.e., the ability to navigate in complicated health care institutions, and to cope with medical information and decisions that have serious consequences. It is likely that most families are dysfunctional in some way, and lack the interpersonal skills, relationships, and conflict management abilities that would make such difficult situations more manageable. Families are tested by the realities of conflicting information about their loved one's condition, the challenge of managing a relative's serious illness while also managing other responsibilities, and the sadness and despair that can often accompany the realization that we are all, after all, mortal.

In Case 3, where young parents have to make decisions for their seriously ill and disfigured child, and in Case 4, where parents have to make decisions for their previously healthy child who has now been diagnosed with a terminal illness, we see many of these factors at work. In these cases a lack of credible information exacerbated problems in decision making. In Case 6, where the mother of a child deemed to be brain dead cannot accept this fact (or her own culpability in her child's death), and in Case 7, where the patient's mother blocks the palliative care team from mentioning how close to death her son is, we see the depths of denial and the extremes to which some family members will resort in order to hear only the information that conforms to their expectations and desired outcomes. Case 10, where the patient's wife pits her husband's oncologists against the intensive care team, is an even more extreme example of someone hearing only what she wishes to hear or to be true.

As a rule, health care professionals cannot simultaneously care for patients and serve as counselors to help families overcome the patterns, habits and beliefs that accompany them to their loved one's bedside. It is likely that there will always be situations such as those that we describe here. Our definition of autonomy and how it is best to be respected do not offer much help, since in theory we talk about one patient and his or her physician engaged in information exchange and decision making, but in practice family members are nearly always involved. The models that teach physicians how to discuss "bad news" with patients make similar assumptions: They encourage physicians to involve family members in decision making but do not provide guidance as to how to manage these difficult interactions. Utilizing the expertise of social workers and palliative care team members with this specialized knowledge could go a long way toward helping, and these professionals should not be blocked from their important role in patient care, even if their presence highlights certain realities such as the patient's impending death. If the question is asked, "should we consult palliative care?" the answer should always be "yes," and anyone involved in the patient's care should be able to request such assistance.

Problems ofHierarchy andTeamwork

When viewed as a whole, the cases in this volume may be seen not only to offer suggestions about what perhaps*should, or*should not*, have happened in the situations presented, but also as sources of practical advice that could be offered to various "branches" of health care comprising the end-of-life tree. Among those separate branches, otherwise identified as particular groups of persons functioning as component parts of the health care process, we would include the following: physicians, nurses, pharmacists, aides, technicians, hospital administrators, hospital ethics committees, social workers, hospice workers, spouses or partners of patients, other family members and friends of patients, and any others necessary for providing the best health care possible for the sick and suffering. The cases in this collection have demonstrated what many persons who have sat with dying loved ones in hospitals know only too well, the tension that exists between physicians and nurses, physicians and families, or physicians and virtually anyone else who might be part of the health care team. Physicians are traditionally respected, deferred to and treated as the pinnacle of responsibility and expertise and with good reason. Nonetheless, the respect accorded to physicians has not infrequently evolved into an unjustified feeling ofintimidation on the part of others. We believe there is simply no reason for the nursing staff in either Case 4 (where a child was denied appropriate pain medication) or Case 10 (where the patient dramatically changed his treatment preferences when his wife was present) to have acquiesced as they did instead of challenging, albeit professionally, decisions or orders they knew had not been made according to their patients' wishes or best interests.

We furthermore believe it imperative that nurses engage responsibly and professionally with physicians in this regard in order to further the well-being of their patients, and we believe this responsibility should be viewed as a moral obligation. If there is to be an ethic of health care designed around the patient's good, it cannot be subject to rank or privilege. Furthermore, if the team concept in health care means anything beyond a nice catchphrase, and we believe it does, it can only be effective where it functions unfettered in an environment consisting of multiple voices that are free, and encouraged, to speak as part of a process, albeit with the understanding there will always be some one person or entity charged with the ultimate responsibility of having to make important decisions, whether this be the patient, the surrogate, the physician, or whoever may be ethically and/or legally appropriate.

Cases 4 and 10 are particularly good examples of situations where nurses became frustrated by the suffering endured by their patients, yet did not feel empowered to speak up to attending physicians, believing that to do so would be insubordinate. Our experience in health care has led us to believe that this widely held attitude must change for the patient's good, and we believe that it is changing, slowly, where there is open communication and especially where physicians are not threatened by respectful input offered by other professionals on the team.

Another type of intimidation that we feel warrants serious attention today is evident in cases where physicians, nurses, entire health care teams, and even hospitals themselves appear to be held hostage by patients and/or families of patients. Case 3 (where young parents must make decisions for their seriously ill newborn), Case 6 (where the patient's mother refuses to acknowledge that her daughter is brain dead), and Case 7 (where the patient's mother refuses to let the palliative care team discuss options for the care of her dying son) are excellent examples of this sort of intimidation. We certainly understand the reluctance of physicians to challenge, argue or "lock horns" with patients and families over treatment decisions, particularly when they have been their patients' primary care providers for some time, perhaps their entire lives. And, we most certainly appreciate the litigious nature of American society today. Nevertheless, we also believe that health care cannot afford to be held hostage to unreasonable demands. Medically appropriate care, humanely and ethically administered, is what is called for, and we trust those qualified to deliver this care will find ways to do so assertively and with the confidence that they are doing so for the ultimate good of patient and family. On the other hand, no one involved in health care should ever attempt to intimidate a patient, even if only because it is thought feasible, such as playing upon the fears of a foreign national unfamiliar with our laws, as evidenced in Case 2, where a Chinese couple needed to make decisions for their Down syndrome infant.

We also believe that there are institutional and organizational opportunities for changes in policy. In our view, whatever good hospital policies may have existed in the cases treated in this book, there was either a lack of appropriate policies designed to alleviate or avoid some of the problems discussed, or little to no effort by hospital administrators to help enforce such policies. We feel strongly that as hospitals undertake periodic review and revision of their policy manuals every effort should be made to ensure that reasonable policies are designed to be flexible, protective of the rights of patients and families, and at the same time protective of hospitals so that the latter do not become victims of bullying at the hands of the former. Case 6 (where the patient's mother refused to acknowledge that her administration of expired insulin had caused her daughter's illness) showed how an overbearing parent could manipulate hospital policies and administrators to her advantage by utilizing local media and legal threats. We believe that hospital administrators and members of hospital ethics committees can play significant roles in this regard; we come to this conclusion based not only upon the evidence found here, but also upon the observations we have each made over years of service on ethics committees.

Last Thoughts

This casebook has introduced numerous situations where things went about as poorly as possible at the end of a person's life. While we hope that these cases are thought provoking, instructive, and a bit unsettling, it is also important to keep in mind that things often go quite "right" at the end of life, where there are many instances of cooperation, dignity,closure, and compassionate care. The ways in which medical schools are educating new physicians, combined with the millennial generation's exposure to vast amounts of information and technology, their experience working in teams from a young age, and passionate idealism might indeed be the perfect combination of skills for a new generation of medical talent.

A palliative care fellow we recently met raised the following situation for discussion at the monthly meeting of an end-of-life research group of which we are long-time members. We were discussing patients' overall reluctance to complete advance directives, and this young doctor added that in her experience even more resistance comes in the form of older physicians more accustomed to dealing with patients in paternalistic ways. The fellow had been picking up additional shifts "moonlighting" at a Veterans Affairs hospital. It was close to midnight, and a 97-year-old man was brought to the facility by ambulance from a nursing home. The man was quite short of breath and likely had pneumonia. He managed to gasp out "please call my daughters in Oklahoma and Wisconsin!" and he produced a piece of paper with their telephone numbers. The fellow conferred with the attending physician, who said, "just intubate him so we can all catch some sleep! You can make your phone calls in the morning." The fellow gave her patient oxygen to ease his breathing and called first one, then the other, of the man's daughters. Both women seemed as though they expected the news they received, as both had been in close touch with their father in recent days. Both women told the fellow that their dad had an advance directive that specified that he did not want to be intubated, and that he wanted comfort measures only. One of the daughters agreed to fax the document to the VA and did so as soon as she finished talking with the fellow. Both daughters agreed that their father's wishes should be respected and he should be allowed to die peacefully. They acknowledged that he had been dealing with numerous health issues in the past year which is what had prompted him to document his treatment preferences and discuss them with his daughters.

Once the document was received, the fellow talked with her patient about his stated treatment preferences and told him that without intubation he would surely die. The man said he knew that was the case and that he was at peace with his decision. He whispered that he had had a very good, long life and was grateful that the fellow took the time to speak with him and his family. The fellow ordered a morphine drip for her patient, who fell asleep and then drifted in and out of consciousness for several hours. The fellow sat by his bedside until he died and she could declare time of death. She then called both daughters to tell them that their dad had died peacefully, and that he was not alone.

The next morning at rounds the fellow came under severe sanctions from the attending physician for the way in which she had handled this patient's situation. He felt she had "wasted everyone's time" and that no harm would have been done had she "followed orders." The fellow went on to tell us that internal medicine residents, and even palliative care fellows, are evaluated based upon their efficiency, ability to adapt to the medical hierarchy (attending physicians first, fellows and others last), and detachment. It is more important to appear business-like than it is to show compassion or to take the time necessary to listen to patient and family concerns. Thefellow also told us that one situation like this, or at most two, would be enough for most of her peers to abandon any commitments to patient-centeredness, empathy and ethically right actions that they might have had prior to medical school and, moreover, that a de-emphasis of these commitments was reinforced through the didactic sessions of medical school.

Now, not all established physicians act in the callous way seen here, but those of us with some experience in end-of-life situations know only too well that putting this patient on a ventilator would serve no medical purpose, and in fact, would have been expressly against the written wishes of this patient. His daughters would then have been faced with the difficult decision of determining if and when their father should be taken off life support, which would have involved coordinating a multi-state trip. And meanwhile, their father would have been in intensive care, tethered to a machine that he had expressly stated he did not want.

It may seem in some cases included in this volume that we are "doctor bashing," but nothing could be further from our intentions. What we are trying to say, among other things, is that it often comes down to one person's willingness to speak the truth as they see it in these complicated cases, accompanied by the courage to act accordingly. The way in which medical care is structured, particularly in hospital settings, is that the doctor is the one with the authority and responsibility to be such a truth-teller. Another message we hope to emphasize is that we feel everyone involved in a dying patient's care should be able to speak their truth and encourage right action on behalf of the patient.

Readers who are drawn to this casebook are the "complicit outsiders" we believe can continue the crucial moral conversation about how best to reform healthcare, promote health literacy around the end of life, and work to make death an accepted and expected part of life. We hope in some measure to have succeeded in providing readers a means of sharpening their own skills in ethical analysis and problem solving through the lessons and perspectives offered here, and that discussing these cases in classrooms, ethics committee meetings, and other places contributes to other on-going efforts to improve the ways people die in America. While most people say they want to die at home, nearly 40% of people die as hospital in-patients (Xu 2016). The hospital setting presents unique challenges for improving end-of-life care—a norm for aggressive rather than palliative care; fragmentation and specialization of care; a hierarchical and institutional setting; conflicts between families, patients, and health care professionals—all of which have been well-documented in the cases presented here. There is no reason, however, why hospitalized patients cannot experience deaths that are more comfortable and peaceful; indeed that happens for the majority of hospitalized patients who die. The cautionary tales and wicked problems discussed here indicate, however, that there is still more work to be done by those of us committed to improving the quality of dying in America.

References

Arnett, Ronald C., Janie M. Harden Fritz, and Leanne M. Bell. 2009. *Communication ethics literacy: Dialogue and difference.* Los Angeles: Sage.

Benhabib, Seyla, and F. Dallmayr (eds.). 1990. *The communicative ethics controversy.* Cambridge: MIT Press.

Cole, Teju. 2017. Getting others right. *New York Times Sunday Magazine*, June 13, 2017. https://www.nytimes.com/2017/06/13/magazine/getting-others-right.html.

Gracia, Diego. 2012. The many faces of autonomy. *Theoretical Medicine and Bioethics* 33: 57–64.

Roscoe, Lori A., Jillian A. Tullis, Judith C. McCaffrey, and Richard R. Reich. 2013. Beyond good intentions and patient perceptions: Competing definitions of effective communication in head and neck cancer care at the end of life. *Health Communication* 28: 183–192.

Xu, Jiang. 2016. *QuickStats*: Percentage Distribution of Deaths, by Place of Death—United States, 2000–2014. MMWR Morbity and Mortality Weekly Report, 65:357.http://dx.doi.org/10. 15585/mmwr.6513a6.

Index

© Springer International Publishing AG 2017
L. A. Roscoe and D. P. Schenck, *Communication and Bioethics at the End of Life*,
https://doi.org/10.1007/978-3-319-70920-8

'

The manufacturer's authorised representative in the EU is Springer
Nature Customer Service Centre GmbH, Europaplatz 3, 69115 Heidelberg,
Germany. If you have any concerns regarding our products, please
contact ProductSafety@springernature.com

Printed and bound by CPI Group (UK) Ltd, Croydon, CR0 4YY
27/04/2026
02097572-0005